Critical Discourse Studies and Technology

ALSO AVAILABLE FROM BLOOMSBURY

Contemporary Critical Discourse Studies, edited by Christopher Hart
and Piotr Cap
Discourse, Grammar and Ideology, Christopher Hart
Introduction to Multimodal Analysis, David Machin
The Bloomsbury Companion to Discourse Analysis, edited by Ken Hyland
and Brian Paltridge
Transductions, Adrian MacKenzie

Critical Discourse Studies and Technology

A Multimodal Approach to Analyzing Technoculture

IAN RODERICK

Bloomsbury Academic
An imprint of Bloomsbury Publishing Plc

B L O O M S B U R Y
LONDON · OXFORD · NEW YORK · NEW DELHI · SYDNEY

Bloomsbury Academic

An imprint of Bloomsbury Publishing Plc

50 Bedford Square
London
WC1B 3DP
UK

1385 Broadway
New York
NY 10018
USA

www.bloomsbury.com

BLOOMSBURY and the Diana logo are trademarks of Bloomsbury Publishing Plc

First published 2016

© Ian Roderick, 2016

Ian Roderick has asserted his right under the Copyright, Designs and Patents Act, 1988, to be identified as the Author of this work.

All rights reserved. No part of this publication may be reproduced or transmitted in any form or by any means, electronic or mechanical, including photocopying, recording, or any information storage or retrieval system, without prior permission in writing from the publishers.

No responsibility for loss caused to any individual or organization acting on or refraining from action as a result of the material in this publication can be accepted by Bloomsbury or the author.

British Library Cataloguing-in-Publication Data
A catalogue record for this book is available from the British Library.

ISBN: HB: 978-1-4725-6949-3
PB: 978-1-4725-6948-6
ePDF: 978-1-4725-6950-9
ePub: 978-1-4725-6951-6

Library of Congress Cataloging-in-Publication Data
Names: Roderick, Ian, 1964- author.
Title: Critical discourse studies and technology: a multimodal approach to analysing technoculture/Ian Roderick.
Description: London, UK; New York, NY, USA: Bloomsbury Academic, an imprint of Bloomsbury Publishing PLc, 2016. | Series: Bloomsbury advances in critical discourse studies | Includes bibliographical references and index.
Identifiers: LCCN 2015040033 | ISBN 9781472569493 (hb) | ISBN 9781472569486 (pb) | ISBN 9781472569509 (epdf) | ISBN 9781472569516 (epub)
Subjects: LCSH: Technology–Philosophy. | Technology–Social aspects. | Rhetoric.
Classification: LCC T14 .R54 2016 | DDC 303.48/3–dc23 LC record available at http://lccn.loc.gov/2015040033

Series: Bloomsbury Advances in Critical Discourse Studies

Typeset by Deanta Global Publishing Services, Chennai, India
Printed and bound in India

For Brynn, Caelan, and Natalie

Contents

Acknowledgments viii

Introduction 1

1 Defining technology: Technology as apparatus 9
2 Multimodal critical discourse analysis 29
3 Analyzing multimodal discourse: A toolkit approach 53
4 Discourses of technology as progress 93
5 Discourses of technological determinism 117
6 Discourses of technological fetishism: (Over)valuing technologies 139
7 Discourses of technological (dis)satisfaction: Consuming technologies 169

Conclusion 193

Notes 199
Glossary 200
Bibliography 207
Index 217

Acknowledgments

I would be remiss if I did not express my thanks and appreciation to a number of people who helped me to write this book. For their generosity and support in all stages of the preparation of this manuscript, I must thank Michał Krzyżanowski, David Machin, and John E. Richardson. For their continued support and patience, I profess my gratitude to Gurdeep Mattu and Andrew Wardell at Bloomsbury. I should also like to acknowledge the work done by the anonymous referees who gave their time and careful consideration to write invaluable responses to the original book proposal. I will also take this opportunity to convey my appreciation of the support and good will that I have received from my colleagues at Wilfrid Laurier University. I must particularly thank and acknowledge those of my colleagues at Laurier who, through the exploitive contracts that they must work under as contract academic faculty, have indirectly subsidized the sabbatical that I took to write this book.

Previously published work included in this volume:

The analysis of the Honda commercial, "Museum" in Chapter 4 and the analysis of the robot-worker commercials in Chapter 7 are derived from Roderick, I., 2013, "Representing robots as living labour in advertisements: The new discourse of worker–employer power relations," *Critical Discourse Studies*, 10(4), pp. 392–405.

The analysis of the representation of EOD robots in Chapter 6 is taken from Roderick, I., 2010, "Considering the fetish value of EOD robots: How robots save lives and sell war," *International Journal of Cultural Studies*, 13(3), pp. 235–53.

Introduction

When discussing the role of technology in their everyday lives, people often talk in terms of its impact. For example, the impact of the printing press, the computer, and the Internet on society have all been heavily discussed in both academic and everyday discussions. The idea is that technology brings with it changes to the way we experience our everyday lives and that our talk of technology reflects those experiences. This is supported by the assumption that the experience of technology is a product of the impact made by the technology itself. However, to talk in terms of impacts means that the technology must enter our lives from somewhere else, already fully formed. Furthermore, it assumes that how we talk about and represent technology is, in turn, a product of those imposed experiences. This book begins with a very different premise. What if we started by assuming instead that the way we experience technology and the way it comes to evolve are actually formed in part by the way that we talk about and represent technology—that the relationship between experience and talk of technology is more synergistic than we might otherwise imagine?

In response to that question, this book looks to challenge our received knowledge about technology and the kinds of technoculture it makes possible. The concept of technoculture is adopted here because it presupposes a relationship between technology and culture in which each element of the relationship is understood as being equally constitutive of the other. Rather than drawing upon discourses of technology that are premised upon dualism, technoculture presupposes that the two are not distinct spheres of activity and knowledge but rather that the one is always implicated in the other. In effect, technology is always already cultural and culture is always already technological. It means that we cannot refer to technology in terms of impact since that would be premised on the idea that there is this separate realm called technology that comes into its own and then imposes itself upon a culture that was previously "untouched." Quite simply, there can be no "outside" of culture and, equally, there can be no "outside" of technology. To borrow from Latour (1986: 2), the divide between the two is merely a border to be "enforced arbitrarily by police and bureaucrats." This has considerable consequences for those who seek refuge in technological determinism. It is, after all, this sort of thinking about technology that spurs on the belief that we can talk about periods of time as being defined by a particular technology as in the case of "the steam age" or "the information age."

At the same time, I am not conjuring up technoculture as a kind of sleight of hand that will allow me to make one kind of determinism disappear, to be replaced by another. There is no one overriding technoculture like some sort of phenomenological bracket. Just as we cannot reduce culture to a single definitive technology, it would be equally inane to try to propose a single definitive technoculture. The actual term, technoculture, was popularized by Penley and Ross (1991) in an anthology of the same title, and Shaw (2008: 4) has since defined the term as "the relationship between technology and culture and the expression of that relationship in patterns of social life, economic structures, politics, art, literature and popular culture." This means that technoculture is not just a description of the interdependence of the two realms of human activity but that it is also its expression or, better yet, realization. Technoculture is therefore simultaneously material and semiotic. Moreover, the definition does not preclude the possibility that the interdependence between the two is always realized as an equitable relationship. Technocratic discourse, such as the one addressed in the discussion of disruptive technology found in Chapter 6, clearly represents a state of affairs in which culture is to be put in the service of technology. I would suggest, therefore, that we think of technoculture as a contested terrain upon which social actors engage in struggles over values, resources, and meanings.

By invoking this image of technology as a contested terrain, I am, of course, calling attention to the ways in which power relations are invested in technology. I do not wish to suggest a simple instrumental notion of power whereby technology gives one group power over another (though power relations between social groups are fraught with asymmetries, and technologies are never "innocent" in terms of reproducing these asymmetries) or, more naively, that technology itself has power over us. Instead, technology is understood as a material confluence of knowledges, practices, beliefs, and expectations that are unevenly distributed among social actors. The development of a technology is never down to serendipitous chance. As Raymond Williams (2003: 7) argued, technologies are always interlocked with "known social needs, purposes and practices." Built into the technology is the social context and, therefore, social relations that support the technology and in that way it can be said to rematerialize and recontextualize those relations and, therefore, the structural asymmetries and inequalities that are already entrenched within the society.

Take, for example, the wearable fitness tracker and the practice of self-tracking. A small, streamlined device typically worn on the wrist, this device allows wearer to generate and collect a variety of biometric data encompassing step counts and distance traveled, levels of physical activity, heart rate, caloric consumption, and even "sleep quality." Looking at the small, aesthetically pleasing bracelet on one's wrist, one might be tempted to imagine it as merely a tool through which one, the wearer, is able to record the bodily movements and blood oxygen saturation levels generated by one's physical activity and translate them into measurements of one's level of fitness. From this perspective, the technology is limited to a device on one's wrist that takes certain inputs and in turn generates desired outputs. And yet, such a perception overlooks the connections that extend beyond the immediate tracker-wearer relation. Instead, we need to highlight how the device is implicated in the generation of new "connections between data, bodies, and self-improvement" (Crawford et al. 2015: 480). For instance, such devices are typically promoted as empowering devices to the user as fitness consumer by virtue of the metrics that they display back to the wearer; however, in doing so, the wearable also "makes the user known to a range of other parties" (Crawford et al. 2015: 480). Furthermore, companies and insurance providers are increasingly making such devices available to insured employees as part of employee "wellness" programs, meaning that the information is also made available to set norms by which individual employees can be measured. Thus, as a form of digital Taylorism, the wearable fitness tracker can be used to expand the measurement of productivity into domains that might once have been considered private and personal through the collection of employee "health" metrics. Accordingly, technology is able to make social relations durable and resistant to change, but, at the same time, as it will be argued in Chapter 1, technologies are never entirely fixed and finalized and so they also carry with them the potential to be reworked and reordered so as to afford new connections between bodies, devices, and society. Understanding technical objects like the self-tracker in this way necessitates a set of critical tools for unpacking how we come to understand the relationship between technology and culture and the structuring of everyday life.

This book is intended for readers interested in the application of discourse analysis to the study of everyday life. More specifically, it offers a systemic approach to analyzing how our understandings of technology and the ways in which we engage with it, are discursively constituted. This entails enlisting a Critical Discourse Studies (CDS) approach that brings together scholarship in Critical Discourse Analysis (CDA), social semiotics, and Foucauldian Discourse Analysis. CDS does not represent a single theory of discourse or method of analyzing discourse but rather an extensive set of critical tools for addressing the relationship between semiosis and social structures and practices. Van Dijk (2009: 62) offers two primary reasons for using CDS over CDA as the broader, blanket label: first, it "suggests that such a critical approach not only involves critical analysis, but also critical theory, as well as critical applications," and second, using the "designation CDS may also avoid the widespread misconception that a critical approach is a method of discourse analysis." In this way, CDS represents a set of tools that researchers can draw upon in order to theorize and analyze the subject of their studies instead of a monolithic method that must be strictly applied in the same way every time.

This is not to suggest that CDS lacks rigor. Inheriting the systemic-functional orientation of Hallidayan linguistics, research practice in the social semiotic approach, for example, entails carefully and systemically detailing how communicators actively draw upon available semiotic resources. As Machin and Mayr (2012: 1) have discerned, "There has been an increased sense of there being value in carrying out more thorough and systematic analysis of language and texts than is permitted through content analysis-type approaches or the more literary style interpretation of Cultural Studies." Doing the work of CDS then requires engaging in analyses drawn from methodical and diligent description in order to support the theoretical claims being made. The analyses offered in this book strive to demonstrate this rigor when applied to discourses of technology.

Historically, research that falls under the rubric of CDS prioritizes "the communication and discursive construction of social, including political, knowledge as well as with linguistic persuasion and manipulation" (Hart 2011: 1). For the most part, research produced by those working within this approach has focused upon the representation of immigrants, the poor and working classes, women, and other marginal or subaltern groups as they appear in journalism, policy documents, and educational materials. In this book I intend to demonstrate how this approach can equally be applied to the study of technology and the ways in which it is organized into the everyday. Of principal concern is how technologies, frequently represented so as to divorce them from their wider contexts of production and use, can only be fully understood in relation to a fabric of social relations that tends to be otherwise suppressed.

The examples that I draw upon illustrate common, received discourses on technology that are well established in the science and technology studies (STS) literature, which in turn informs Chapter 1. As such, my objective is not to prove the existence of these discourses but rather to demonstrate how they are realized multimodally. To this end, I have selected as examples well-recognized texts that have appeared in a variety of media including magazines, television, newspapers, social media, websites, video games, and newspapers. Accordingly, the texts analyzed in this book are representative of four predominant discourses on technology: technology as progress, technological determinism, technological fetishism, and technological (dis) satisfaction. By engaging in a critical multimodal analysis of these texts, I strive to establish how such texts make tacit normative claims about the "nature" of technology and their users through the use of multiple semiotic modes such as language, images, typeface, and music.

Although CDS scholars have not produced a great deal of commentary on technology and society, there are some precedents for applying it to this task. The collection, *Discourse and Technology: multimodal discourse analysis* edited by LeVine and Scollon (2004), offers a remarkable compilation of diverse essays, but by primarily focusing upon the "technologizing of discourse" these essays tend to focus upon the impact of technology upon discourse, thus treating technology as something external to and impinging upon discourse rather than being an integral part of it. Other more recent work in this vein would be the work by van Leeuwen, Zhao, and Djonov (2014) on what they refer to as semiotic technologies. These essays offer an insightful account of the affordances of media such as PowerPoint, but again the focus is more upon the effects of technology upon discourse. Another related area of study that has been touched upon by CDS scholarship is the communication of science and technology. The collection, *Reading Science: Critical and Functional Perspectives on Discourses of Science*, edited by Martin and Veel (1998), and the work of Myers (1996, 2003) exploring Actor-Network Theory in relation to CDA offer interesting discussions of how science is communicated to the public but more specifically in the context of pedagogic practice and risk communication and scientific popularizers, respectively.

Although the study of technology and the everyday has not featured prominently in CDS, this does not mean that it is beyond its purview. CDS is a political practice as much as it is an academic one. Practitioners see their work as explicitly political and strive to adopt an advocacy role. While technologies are often thought of as being simply practical and useful things, science and technology scholars make it clear that technologies are not neutral and that they do in fact materialize politics. For example, Winner (1980: 124) argues in his well-known essay "Do Artifacts Have Politics?" that embedded in the design of highway overpasses was a politics of exclusion that prevented buses from using the parkways in Long Island, NY, and thus "poor people and blacks, who normally used public transit, were kept off the roads." Likewise, Latour (1990) argues that technologies are "society made durable." Finally, Slack and Wise (2005: 2) see technologies as occupying "sites of struggle over meanings and power, and that they can both reinforce and undermine structures of inequality." A CDS approach to technology, therefore, is not simply interested in representations of technology—looking to separate those that distort from those that accurately represent technology. Rather, it is very much a political engagement with the politics of technology, where they are practiced—namely, the everyday.

A brief comment on terminology

The unavoidable use of the terms CDS, CDA, and Multimodal Critical Discourse Analysis (MCDA) and switching between them does leave open the potential for confusion. I shall strive to use the term CDS inclusively when discussing critical approaches to discourse analysis as a whole and referring to CDA to either reflect authors' self-descriptions of their work or to indicate work that is more explicitly based in critical linguistics. Similarly, I will use social semiotics to refer to that body of work directly inspired by Kress and van Leeuwen (2006), Kress (2010), and van Leeuwen (2005) that focuses upon semiotic modes and resources. MCDA is understood here more as a practice than a body of scholarship like CDA or social semiotics. I also use the term to distinguish it from the more technically focused Multimodal Discourse Analysis (MDA).

Chapter contents

Chapter 1. This chapter introduces readers to the theory of technology that shall underwrite the analyses to follow in the subsequent chapters of this book. If we are to have an alternative to a-critical theories of technology, the starting point is not only to negate those theories but also to begin to build a theory (or set of theories) that can afford different ways of knowing and engaging with technology. The chapter begins with the analysis of a magazine advertisement that reproduces the technology-culture opposition that serves to structure how we typically imagine our relations with technology. The work of Simondon is then introduced as the impetus for this rethinking of the relation between technology and culture. Having offered a summary of Simondon's non-hylomorphic account of the relationship between technology and culture, the chapter concludes by proposing that technologies would be better understood as apparatuses rather than as simply useful tools.

Chapter 2. This chapter introduces the theoretical framework for CDS. The chapter contextualizes the concept of discourse explaining its linguistic and non-linguistic uses. Discourse, it will be argued, functions as a resource for constituting what can be known about a particular subject. The concept of semiotic resource is then developed. Mode is then introduced to highlight how semiotics such as speech, dance, music, and so on function both as resources for communication and as meta-resources for accomplishing representational, interpersonal, and compositional communicative functions. At the same time, it is noted that communicative events rarely entail the use of only one semiotic mode and so the theory of multimodality is then introduced to address the way in which meaning is produced recombinantly across semiotic modes. Finally, Iedema's (2001: 36; 2005) conceptualization of the process of resemiotization is presented.

Chapter 3. This chapter lays out the methodological approach that will be adopted in order to develop a critical study of technological discourses and their articulation within our everyday technocultures. It begins by addressing the matter of being critical within CDS practice and situates it in relation to Latour's notions of "matters of fact" and "matters of concern." In addition to the Frankfurt School, CDS draws directly from Foucault's own theorization of discourse and his conception of visibility is presented as a solution to the problem of tacitly inheriting a kind of camera-obscura epistemology in which ideology is understood as the world turned upside-down. The chapter then offers a rejection of the supposed contrariety between description and critique, arguing that description can be enlisted as a tool of critique when the two are applied sequentially. This leads to adopting Machin's proposal for a toolkit approach to CDS.

Chapter 4. This chapter is about the progress narrative. It starts by examining how progress was constituted as a measure of and a means to moral and social betterment. As technology also became a measure of moral and social superiority, it became attached to the progress narrative. Initially part of grand narratives of nationalism in the nineteenth century, progress came to be increasingly more mundane and something to be experienced in people's everyday lives, for better or worse. This is demonstrated through an analysis of Carousel of Progress at Disney World of a commercial promoting Honda using its Asimo robot.

Chapter 5. This chapter is about technological determinism. Technological determinist discourse attributes to technology the ability to change society for the better or for the worse. Accordingly, examples of the way in which mobile technologies and social media are depicted as undermining social relations are used to demonstrate how technology can be constituted as a source of destructive societal change. The chapter then turns to the marketing of the iPhone and the so-called "disruptive technology" in order to explore how technology can just as readily be represented as a revolutionary source of societal change. In each case, social change is presented as being brought on by the introduction of a technology as if it suddenly just appeared causa sui.

Chapter 6. This chapter deals with technological fetishism. It starts by considering how the experience of the sublime was attributed to technology. The technological sublime is closely related to technological fetishism but like progress, it does not hold the same currency since technologies themselves have come to be experienced as more mundane and everyday than compared to those of the nineteenth century. As a result, the sublime itself has come to be secularized as demonstrated in an analysis of the opening to the US Army video game, *Future Force Company Commander*. The chapter then uses the example of Explosive Ordnance Disposal robots to explore how the discourse of technological fetishism, as it is realized, extends beyond technological determinism through the overvaluing of technological objects.

Chapter 7. This chapter addresses discourses of technological consumption and the satisfactions they promise. By problematizing the notion that technologies afford us labor savings and convenience, the chapter argues that the typical way that we experience technology continues to render it as black-boxed. Using the example of the Project Tomorrow after-show, it is argued that technology has been rearticulated from a civilizing force to a source of personal satisfaction. This notion of personal satisfaction and convenience is further discussed through the example of a promotional campaign for domestic vacuum robots and, finally, through a discussion of the representation of robots as workers.

1

Defining technology: Technology as apparatus

The typical dictionary definition of the word "technology" is the application of scientific knowledge to practical purposes. This emphasis upon application and "the practical arts" is largely how technology has been defined since the nineteenth century (Williams 1985: 315). Such definitions reflect two concurrent ways of thinking about technology: as both a body of knowledge and the application of that knowledge. In the first case, technology is understood as practical and applicable knowledge and in the second, technology is understood as a grouping of techniques that make scientific knowledge practicable. In either case, though, as knowledge or as practices, technology is understood to be simply a pragmatic means to a utilitarian end. What such a definition does, in effect, is to conceptualize technology as, simultaneously, allowing us to do things and make changes in the world and, at the same time, believe that the tools that we use to make these changes are somehow politically neutral. As Dumouchel (1992: 409) characterizes this viewpoint, technology is regarded as objectively "applied science, and technical objects [as] wholly transparent artifacts whose sole reality is in the design and intention of those who conceive them." So while etymologically, technology entails both knowledge and the enactment of that knowledge, in everyday parlance, technology tends to connote the objects that materialize how those knowledges are practically applied.

As Slack and Wise (2005: 95) posit, when talking about technology "in popular discourse, however, it is almost always as things." In this way, we are encouraged to think of technology or, rather, technologies as intrinsically neutral tools; that what matters is putting them to the correctly chosen ends and that things-cum-technology are the direct material manifestation of rational thought. In other words, from this perspective, the question of technology is being reductively approached "through the concept of use"

(Dumouchel 1992: 409). Furthermore, such an understanding of technology tends to treat technical objects not only as instrumental means to rational ends, but also as discrete and bounded things isolated from their actual conditions of use. Ironically, what eludes such understandings of technology-as-thing is the very process through which the thingness of technology is realized. What I propose to introduce here is another line of thought regarding technology that offers a very different understanding in which it is approached not in terms of use but rather its becoming and the conditions under which that becoming takes place: in a word, its individuation.

This chapter presents the theory of technology that will underwrite the critiques of this book. The chapter starts by analyzing a magazine advertisement that exemplifies conventional understanding of the relationship between technology and culture. Through a demonstration of how the advertisement presupposes a binary relationship between the two, the possibility of an alternative conceptualization of interplay and traffic between technology and culture is signaled. The work of Simondon is then introduced as the impetus for this rethinking of technology and culture. Having offered a summary of Simondon's non-hylomorphic account of the relationship between technology and culture, the chapter concludes by proposing that a technology would be better understood as an apparatus rather than as simply a useful tool.

The dualist view

Conventional thinking and talk about the relationship between technology and culture has tended to assume that one impacts upon the other. This is based on the assumption that there is a dualistic relationship between the two and that they can be treated as separate and distinct spheres of human activity.

The ad image portrays a man and woman standing face-forward in a kitchen. The two people appear to be standing to attention with linked arms and not actually engaged in any action besides posing for the camera/viewer. In terms of representation of social action, this can be interpreted as a conceptual rather than a narrative-type image. Each communicative event has specific ways of representing participants according to the type of process being realized. Kress and van Leeuwen (2006) propose that most, if not all, visual representations can be divided into either narrative or conceptual types of processes. While narrative representations depict social actors and action, conceptual representations depict social actors in terms of attributes. Narrative processes would therefore tend to include participants undergoing or performing some sort of action, while conceptual processes tend to offer analytical representations of participants. Thus, while narrative visual

Figure 1.1 *This ad for a hybrid oven exemplifies the dualist or hylomorphic view of culture and technology.*

representations "serve to present unfolding actions and events, processes of change, [and/or] transitory spatial arrangements," conceptual visual representations, instead, depict "participants in terms of their class, structure or meaning" (Kress and van Leeuwen 2006: 59). I would argue that the image presents the two figures, standing as they are, in terms of attributes rather than actions. They function as what Machin (2011: 27) terms "carriers of connotation." In this way, the image can be said to realize a conceptual visual representation structure rather than a narrative one.

Looking at the two figures in terms of attributes, we see that they differ in interesting ways. On the left is a man in chef's attire presenting a crown rack-of-lamb roast and on the right is a woman dressed in a silver body suit and holding a silver motorcycle helmet at her side. The man stands with feet together while the woman stands with feet apart. It should also be noted that the man is white while the woman is of Asian descent.

The setting the two occupy is that of a kitchen. Considering the size and finishing features of the kitchen, it is obviously a "high-end" or "upmarket" one. The fact that it is an "aspirational" kitchen should not be surprising since this is an advertisement for the more expensive end of an appliance line. The objects that sit on the counter and the shelves are stylish, modern, glass and stainless steel items. The shelving and wall panel system along with the pendant lights are also of stainless steel. On the left-hand side are primarily small appliances, containers, and vessels while on the right are mostly prepared foods. The kitchen itself is a blend of organic and inorganic materials. On the one hand, there are the presented fruits, vegetables, and prepared foods as well as the wooden cabinetry. On the other hand, there are the industrial material finishes such as stainless steel, glass, and acrylic. The orange-brown cabinetry brings some warmth to the room and balances the coldness of the white panels, stainless steel, and the polished floor surface. At the same time, the flat, smooth surfaces with no visible joinery preserve the modern style. The clear acrylic kitchen chairs and stainless steel are reminiscent of mid-century modern and also contribute to the modern, high-tech aesthetic. Likewise, the stainless steel finishes on the cabinetry appear to be burnished so as to resemble solid-state circuit boards, further adding to the high-tech sensibility.

Interpersonally, the dyads occupy a position of social distance since they are presented in a full-body shot. The horizontal angle is fully frontal, which as it will be addressed in Chapter 3, has the potential to signify a high degree of involvement between viewer and subjects. At the same time, the vertical angle of interaction is lowered so that the viewer looks upwards even though the two stand as if awaiting the viewer's inspection. This would suggest that despite their being offered to the viewer, the two should still be held in esteem.

Because the primary representational function of the image is the symbolic representation of the participants rather than their actions, each figure functions as the embodiment of a particular concept. The anchoring text, which appears at the top of the ad, reads "Superb marries supersonic." With their arms linked, the couple is the visualization of superb marrying supersonic. The silver, streamlined attire and helmet carried by the woman would suggest that she is supersonic. Her body suit, with the footwear worn inside the leggings to make it seamless, looks almost liquid like quicksilver or mercury. The man, on the other hand, not just someone who cooks but a chef, holding the perfectly presented crown rack, is the personification of superb. The dyads also not so subtly play to racial stereotypes of whiteness being associated with civilization and culture while Eastern Asia is associated with consumer culture and technology. If he represents mastery of cuisine and culture, she represents the aesthetics of speed and technology—indeed, she appears to be the tool/appliance to which his knowledge/mastery can be applied. In sum, this is an image that conceptualizes the relationship between technology and culture as a joining of two distinct spheres much as heterosexual marriage has been conceptualized in hetero-normative patriarchal discourse.

The compositional elements of the advertisement also contribute to this representation of culture and technology as two binary opposites that are then wedded together. The advertisement is essentially divided into three vertically stacked sections in which the kitchen image appears in the middle. The top and bottom portions appear as white space and the image is bordered by gray text that echoes the stainless steel of the kitchen. Different features of the oven are listed in the text. The lettering creates a broken rather than a solid border so that rather than neatly separating the top and bottom text from the image, the border seems more porous. Being the color gray, it actually brings the gray of the stainless steel in the image into the two white panels. At the same time, the significantly raised initial letter S in Superb is colored the same orange-brown as the wood front panels of the kitchen cabinets, further adding to the visual cohesion of the advertisement but also further linking "superb" to the organic elements of the kitchen. This color is also used in the bottom text panel for the text that explains how the oven actually works. The choice of this color balances and creates cohesion between all three sections of the ad since it appears in all three sections with the S in the top section, the cabinetry in the second, and the two blocks of text in the bottom section. In the third panel, though, the color is no longer tied exclusively to Superb since it is now coloring text describing the technical details of the appliance. Using the orange-brown to color the "real" of the appliance introduces another redundancy to the advertisement since it is also a joining of the color coding of superb with the features of supersonic.

The logic of the advertisement is therefore premised upon a notion of technology and culture as being two separate spheres that are then brought together. Each sphere has its own distinctive properties and responsibilities. Culture is aesthetic while technology is utilitarian. Culture is slow while technology is speed. The oven itself is the synthesis of this dialectic and becomes a device for inputs and outputs. In goes the efforts of a good cook and out quickly comes the work of a great chef. As I shall discuss below, this precisely represents the hylomorphic view of the relationship of culture and technology.

The individuation of technical objects

Gilbert Simondon was a French philosopher who has only recently begun to be read widely by English language scholars. He defended both his principal doctoral dissertation, entitled *Individuation in the light of the notions of Form and Information,* and his secondary thesis, entitled *On the mode of existence of technical objects,* in 1958. His theory of individuation is highly influential and has directly influenced the work of philosophers like Gilles Deleuze, Bruno Latour, Bernard Stiegler, Andrew Feenberg, Isabelle Stengers, and Brian Massumi. Though Simondon is not exclusively a philosopher of technology, his conceptualization of individuation, transduction, and associated milieu is such that it

> separates him from other well-known thinkers of technology, such as Martin Heidegger and Jacques Ellul. Rather than looking for the "essence" of technology (Heidegger) or the formal logical relation of technics (Ellul), Simondon repeatedly draws upon details from the historical development of technologies in order to show that it is in the concrete instantiation of specific technical objects that one finds the key to understanding the ontology of technics. (Hayward and Thibault 2013: 30)

In this way, looking at technical objects as being subject to a process of individuation rather than as being already finalized forms, we can begin to develop a theory of technology that will aid us in our efforts to, borrowing from Wodak and Meyer (2009: 7), make visible the interconnectedness of technocultural things.

Instead of understanding the technical object as having arrived before us in an already finalized form, Simondon asks that we consider it as something in process: "Every technical object undergoes a genesis" such that the "unity, individuality, and specificity of a technical object are those characteristics which are consistent and convergent with its genesis" (Simondon 1980: 18). What gives the object its specificity, its "being," is already implicated in its "becoming." The "thingness" of the technical object is therefore not a priori to its existence in an environment but, at the same time, the environment is always already constituted in and through the object. As Del Lucchese (2009: 181) succinctly puts it, "Being is not what 'is'. . . . Being is what becomes in and through relationality." Technological objects, in this way, are not defined in terms of intrinsic properties but rather are constituted as discrete entities only insofar as they can be individuated, and this is always accomplished relationally between object and environment.

Individuation is the term Simondon uses to refer to the process that brings about the genesis of the technical object. Individuation is not to be confused with individualization and is meant to suggest not the development of a finite and finalized entity but rather the generation of a provisional individual in an ongoing state of becoming. The potential to become is never exhausted by individuation, so there is always the possibility of further individuation, of becoming differently. What this means is that individuation has no implicit final end-state as individualization does. Individuation constitutes a break with hylomorphic thinking since it does not imply a process whereby matter takes on a predefined form but rather one where form is always becoming. Individualization, under hylomorphism, leads to developing a predefined final individual form; individuation, in contrast, entails an emergent conception of form that is never finalized and always taking form.

It is Simondon's example of brick-making that makes clear how individuation breaks from hylomorphism. As Ingold (2012: 433) points out, the example is particularly well chosen since it lends itself so readily to the hylomorphic understanding of the relationship between form and matter. From a hylomorphic perspective, the clay is unformed matter and the brick mold determines the final shape that the clay will take in order to form a proper brick. But, as Ingold (2012: 433) elaborates, this perspective depends upon a number of misperceptions:

> For one thing, the mold is no geometric abstraction but a solid construction that has first to be carpentered from hardwood. For another thing, the clay is not raw. Having been dug out from beneath the topsoil, it has first to be ground, sieved, and kneaded before it is ready for use. In the molding of a brick, then, form is not united with substance. Rather, there is a convergence of two "transformational half-chains" (demi-chaînes de transformations)—respectively, constructing the mold and preparing the clay—to a point at which they reach a certain compatibility: The clay can take to the mold and the mold the clay.

The brick then is formed not simply out of a mold but rather "the contraposition of equal and opposed forces immanent in both clay and mold" (Ingold 2012: 433). Individuation then, unlike developmental models of individualization, constitutes an overcoming of the hylomorphic binary of matter and form (see also Iliadis 2013). To Simondon, accepting the hylomorphic account depends upon reducing brick-making to inputs and outputs and failing to consider the actual process of producing the bricks and, ultimately, the clay-mold relation. The brick-making process has been effectively black-boxed (Latour 1987) such that only the device itself is made to be visible.

The conversion of the clay-mold into a brick is an example of a transductive process. Transduction is the transformational process that underwrites individuation whereby differing domains are brought together as a synergy but not actually a synthesis (see Simondon 1992: 315). As Mackenzie (2005: 393) summarizes, transduction "leads to individuated beings, such as things, gadgets, organisms, machines, self and society, which could be the object of knowledge." In the case of the brick, it represents the converging of different domains: quarry, clay refinement, mold construction, piping, kiln, etc. By proposing that technological objects undergo a process of transduction, Simondon is attempting to alert us to the conditions under which technical artifacts come into existence and how those conditions are both a part of the object's being and themselves transformed by the emergence of the object. The technical object comes into existence or becomes "as a specific type that is arrived at the end of a convergent series" (Simondon 1980: 18). Such a series is understood by Simondon as a movement from an abstract to a more concrete mode of existence whereby "the technical being becomes a system that is entirely coherent with itself and entirely unified." Simondon thus conceives the genesis of the object to be the product of a process of concretization.

Simondon uses concretization to indicate a provisional differentiation and stability that the technical object realizes in the ongoing process of individuation. As Iliadis (2013: 15; see also de Vries 2008: 25) points out, Simondon uses the word "concrétisation" to indicate "an indefinite process that does not indicate a 'transfer' as if something had gone from one state (abstract) to the next (physical)." In this way, the concretization of the technical object does not suggest that it has achieved a final material form, that an inventor's dream has now become reality, but rather that as objects-in-process, technical objects will achieve a contingent and metastable but not finished or finalized form. Accordingly, concretization and the corresponding object are understood more as event than product. Grove (2014) demonstrates this in his study of Improvised Explosive Devices (IEDs), arguing that IEDs must be understood as assemblies that are constantly adapted and adjusted to the milieu in which they are deployed. Furthermore, when looking at IEDs in terms of use, it is easy to imagine them to be a single class of thing; however, as Grove (2014: 14) observes, "Like Darwin's finches, even within this narrow temporal and geographic corner of warfare there are varied attributes and different morphogenetic histories for each subspecies of IED." So rather than the end point of a process, concretizations should instead be thought of as iterations in a series of becomings, with each becoming introducing the potential for yet another new becoming. Each iteration is thus a becoming within its own unique context or associated milieu, so our concern is not with *how* the object came to be but rather the unfolding events of becoming. In the case of the IED, as a particular iteration of an IED variant comes to be reliably defeated by military forces, new iterations are generated by insurgent forces, thus bringing further change to the theater of operation. Thus an IED, as technical object, is always becoming and never finalized and, accordingly, always represents the possibility of further mutations or becomings. The IED encountered on the roadside carries with it the potential for a new IED and therefore is also an IED-in-becoming. Through individuation, therefore, technical objects are concretized as individuated entities and thus are both constituted as part of, but also differentiated from a lineage of, antecedent technical objects. Accordingly, concretization generates object-events that emerge from technical lineages, and as I will argue below, these object-events are also discursive.

The individuation of any technology is therefore realized as a series of refinements that are brought together to form a lineage. Technical lineages are lines of technological refinement that represent the movement from the abstract to the concrete. Each lineage is the path an individuating technology takes from its abstract form into a coherent system in which disparate elements or subsystems have been unified and made to work in an interoperable manner. The lineage is therefore a tracing of becomings or optimizations because with each (re)iteration of the technical object, it becomes more finely resolved to the environment (which is also being transformed by the transforming object) in which it operates. Accordingly, a lineage should not be misconstrued as a linear progression from idea to final development. As Simondon (1980) demonstrates with the vacuum or electronic tube, there are many parallel developments and false turns such that no one singular path leading to the vacuum tube was perfected. Simondon is endeavoring to demonstrate that technical objects evolve in ways akin to biological beings and so a lineage, very much like what Deleuze and Guattari term a machinic phylum, is part of the project of "supplying a microbiology, a morphology, and an ecology of machines" (Lister et al. 2008: 386). Simondon is also quite clear that lineages are to be understood as technical rather than based in use such that the automobile cannot be treated as being of the same lineage as the horse and buggy. The steam engine, the electric engine, and the diesel engine may have all been enlisted to move trains along tracks, but each engine has its own distinctive technical mechanisms and resolved its own specific technical problems in order to be properly put to use. From Simondon's perspective, lineages, accordingly, are strictly defined by the development of the object's own technical features but these features are always developed in the context of the object's environment.

The associated milieu

Simondon refers to the environment to which the technical object is adapted as a milieu. Different kinds of milieus will play host to different kinds of technical objects. To illustrate this, Simondon offers the example of the factory and the locomotive engine: "Because of its singleness of milieu, the factory engine does not have to be adapted to its environment, whereas the traction engine needs an environment of adaptation, which is composed of repressors located in the electrical sub-station or on the locomotive itself" (Simondon 1980: 46). So every technical object emerges and must be adapted within a milieu but the relationship between the two is not one of simple determination. The environment or milieu is not simply the stage on which the technology acts. Simondon (1980: 49) is quite clear that the milieu is never truly distinct from the technology: "This environment, which is at the same time natural and technical, can be called the associated milieu." Thus, there is no presumption of an originary, natural environment that must then bear the impact of a new technology. Simondon's account of the relationship between technology and the milieu is one of mutual co-determination but it does not slip into dualism. Instead, the technology and environment are always already paired to one another. The technology becomes in a particular environment to which it is suited and at the same time the milieu also becomes as it is associated with the technology. It is not possible to erroneously put the cart before the horse because, in essence, Simondon refuses to see the cart and the horse as separate and distinct entities. Instead, he can only see them as becoming horse and cart through a relation: "What does the appearing in individuation is not only the individual, but the individual-milieu couple" (Simondon in Lefebvre 2011: 3). The horse becomes a cart-pulling horse by virtue of the cart and the cart becomes a horse-drawn vehicle by virtue of its being fixed to a horse. In this way, the horse and the cart *become* the horse and cart in relation to one another. In other words, with individuation, both object and milieu are always concretized as a paired relation and never as distinct elements or components.

One limitation that needs to be addressed in Simondon's thinking is that concretization, the milieu, and therefore lineages, as he describes them, are all limited to the purely technical. If the genesis of the technical object is realized through a process of refinement, according to Simondon, the "areas of most active progress are those in which technical conditions outweigh economic conditions" (Simondon 1980: 23–24). Other spheres of human activity are, oddly, separated from technical matters. Social, cultural, and economic domains, all which Simondon reduces to the economic, are the source of distorting "social myths and opinion-fads" that obscure the technical object and cause it to be "not appreciated in itself":

> Economic causes, then, are not pure; they involve a diffuse network of motivations and preferences which qualify and even reverse them (e.g. the taste for luxury, the desire for novelty which is so evident among consumers, and commercial propaganda). This is so much the case that

certain tendencies towards complication come to light in areas where the technical object is known through social myths and opinion-fads and is not appreciated in itself. (Simondon 1980: 24)

Thus, in its strictest sense, transduction is the realization of purely technical solutions to purely technical problems. Citing the added "complication" of power-steering in automobiles, Simondon (1980: 24) declares, in a statement redolent of eugenics discourse, "The automobile, this technical object that is so charged with psychic and social implications, is not suitable for technical progress." It would seem that to Simondon, the concretization of the automobile has been contaminated by forces extraneous to its technological development.

Simondon prefers to demarcate the addition of electrical starter systems, power steering and other such capacities as the "faddish" product of sales promotion and distinct from the technicality of the automobile itself. In part, this rather restrictive perspective can be explained by noting that Simondon is attempting to address what he regards to be a general failure to acknowledge the agency of the technical in human affairs. Rather than claiming that technology is an outside alienating force that has come to dominate culture, which is characteristic of critical-humanist accounts such as those of Ellul (1964), Simondon (1980: 11), in fact, argues the reverse: "Culture fails to take into account that in technical reality there is a human reality, and that, if it is fully to play its role, culture must come to terms with technical entities as part of its body of knowledge and values." Simondon is pointing to a tendency to overlook the role played by technology in human affairs, resulting in an impoverished understanding not only of technology but of society as a whole, much as Latour (1992) has more recently done in "Where Are the Missing Masses?" Thus, I would argue that Simondon's seeming refusal to address the cultural and his narrow elucidation of the milieu needs to be considered in this context. Simondon, in actuality, is not trying to invert the cultural-technology division but instead formulate "a social pedagogy of technics aimed at the reintegration of technology into culture" (Bardin and Menegalle 2015: 15). With this agenda in mind, we can adopt Mills' (2011) position that "it is not just technical developments which are involved in the concretization process but that we can discern other processes/forces involved, such as cultural, economic, social and material which also become concretized in any technical development," and in this way, concretization is to be understood as incorporating both the material and the discursive.

In the process of concretization, accordingly, the existing milieu is transformed and a new milieu associated with the newly emergent object is created. The associated milieu is paired to the object and, as such, the two cannot be separated. They, in effect, co-determine one another. A good example of this would be the "at-once mobile and home-centred way of living" that Raymond Williams (1975: 19) characterizes as mobile privatization. This mode of living comes into existence through not only the relatively affordable family automobile but also the rapid suburbanization during the post-Second World War period in North America. The automobile in one sense creates its own environment or associated milieu of extensive road infrastructure, larger homes with driveways located well beyond walking distance from traditional centers of employment, concentrated retail spaces in the form of shopping malls that service large geographic areas, refrigeration devices that support shopping habits that entail less frequent visits but purchasing in greater volumes, regular and well-placed service stations, and so on. But on the other hand, the "democratization" of the automobile depends upon the creation of an environment that is conducive to the mass use, servicing, and storage of the automobile and so makes the automobile and its particular functionalities (family transportation and cartage) possible. So while one might be tempted to argue that the car brings about these changes, at the same time it is equally valid that these changes were necessary for the mass adoption of the private family car. Rather than trying to impose an instrumental causal chain, the concept of associated milieu affords a more nuanced relational means of interpreting the interaction between a technology and its technological-cultural environment such as the car-mobility relation.

The drone as a technical object

It is precisely the predilection for trying to define technical objects in terms of use (Dumouchel 1992: 409) that leads to imagined lineages that foster understandings of technical objects as utilitarian black boxes that are the result of linear progressions in technical know-how rather than lineages of individuations that are the product of environments arising from complex interactions between technology, culture, and politics. This can be seen in the way military drone aircraft or Unmanned Aerial Vehicles (UAVs) have come to be represented in news media and nonfiction entertainment. Histories of the military UAV typically provide lines of descent that span early military balloon use in the US Civil War (and sometimes the earlier balloon use by Austria against the city of Venice in the First Italian War of Independence) to the present-day use of armed UAVs such as the MQ-1 Predator and the MQ-9 Reaper, both by General Atomics Aeronautical Systems. For example, "A Brief History of Unmanned Aircraft: from bomb-bearing balloons to the Global Hawk" published on airspacemag.com, a publication of the Smithsonian Institution's National Air and Space Museum, presents a history of the UAV, spanning from the US Civil War through to the present, as a series of proto-UAVs leading up to the Northrup Grumman RQ-4 Global Hawk (Darack 2011). The problem with such accounts is that they again foster the image of a linear progression or evolution that was set in motion in the past and incorporates each artifact in terms of use independent of its associated milieu. While a history of the militarization of "airspace" is certainly not without value, there is a price then to be paid (ultimately, in human lives) for thinking of the drone in terms of use alone.

A clear example of this imagined genealogy is a Department of Defense News article that was written in 2002 when the public was just beginning to be interested in drones and just before the use of drones for Central Intelligence Agency (CIA)-overseen "targeted killing" became an established practice in 2004. Presenting it as "From US Civil War to Afghanistan: A Short History of UAVs," the author offers the following timeline for the development of US military UAVs:

> During the American Civil War, both sides tried to use *rudimentary* unmanned aerial vehicles.
>
> According to Dyke Weatherington, deputy of the Defense UAV Office, Union and Confederate forces launched balloons loaded with explosive devices. The [idea], he said, was for the balloons to come down inside a supply or ammunition depot and explode. "It wasn't terribly effective," he said during a recent interview.
>
> The Japanese tried a similar ploy late in World War II. They launched balloon bombs laden with incendiary and other explosives. The *theory* was high-altitude winds would carry the balloons over the United States, where the bombs would start forest fires and cause panic and mayhem. The Japanese weren't able to gauge their success and so called it a flop and quit after about a month.
>
> The United States also tried a type of UAV during World War II called Operation Aphrodite. "There were some *rudimentary* attempts to use manned aircraft in an unmanned role. The limitation there was, we didn't have the technology to launch these systems on their own and control them" Weatherington said.
>
> Allied forces used the modified manned aircraft basically as cruise missiles. The *idea* was a pilot would take off, get the plane to altitude, ensure it was stable and then pass control to another aircraft through a radio link before bailing out.
>
> ...

During the Vietnam War, technology started to make UAVs *more effective*. Weatherington said they were used *fairly extensively* and were *called drones*.

Large numbers of modified Firebee drones overflew North Vietnam. The aircraft, about the size of today's Predator UAV, launched first for simple day reconnaissance missions at varying altitude levels. "They had conventional cameras in them," Weatherington said. "Later on, they were used for other missions such as night photo, comint and elint, leaflet dropping and surface-to-air missile (SAM) radar detection, location and identification."

One of these Firebees hangs in the Air Force Museum at Wright-Patterson Air Force Base, Ohio, amassing over 65 individual missions. As a whole, Firebees flew over 3,400 sorties during the Vietnam War.

Several of the UAVs we know today owe much to Israel, which develops UAVs aggressively. The US Hunter and Pioneer UAVs are *direct derivatives* of Israeli systems, Weatherington said.

The Navy and Marine Corps operate the Pioneer UAV system has been in operation since 1985. Once during Desert Storm, Iraqi troops actually surrendered to a Pioneer.

At the time, the battleship USS Missouri used its Pioneer to spot for its 16-inch main guns and devastate the defenses of Faylaka Island, which is off the Kuwaiti coast near Kuwait City.

Shortly after, while still over the horizon and invisible to the defenders, the USS Wisconsin deliberately flew its Pioneer low over Faylaka Island. When the Iraqi defenders heard the sound of the UAV's two-cycle engine, they knew they were targeted for more naval shelling. The Iraqis signaled surrender by waving handkerchiefs, undershirts and bed sheets.

Following the Gulf War, military officials *recognized the worth* of the unmanned systems. The Predator *started life* as an Advanced Concept Technology Demonstration project. The program hurried the development of the Predator along, and it *demonstrated its worth* in the skies over the Balkans. (Garamone 2002)

Clearly, this is a story of technological evolution that begins with an idea poorly executed using balloons in the US Civil War and later in Second World War and ends with the development of the Predator. Between The Civil War and Second World War, the narrative moves from "idea" and "theory" with "rudimentary" objects that were still more idea than object. With the Vietnam War, the military UAV begins to take form, being "more effective," "used fairly extensively" and "called drones." We are then told that several contemporary UAVs are drawn from Israeli drone technology and two are direct "derivatives" or descendants. The military UAV is then made more finalized or solid when it is reported that, in the Gulf War, Iraqi soldiers once surrendered to a Naval UAV. It is then after the Gulf War that the value of the military UAV is fully recognized and the Predator drone comes to "life" and is made to fly as a fully formed military object.

This, of course, is a rather tenuous "lineage" since, even functionally, the balloon "ancestors" of the Predator are actually munitions rather than a remote weapons platform. The problem with representing the drone in this manner is that it naturalizes and normalizes what is, in fact, the result of historical decisions to allocate resources and engage in warfare in particular ways. The armed drone aircraft and its use is made to seem inevitable when in reality, the drone came together via a series of deliberate decisions made under very specific geo-political conditions: "The Predator is a creature born of the War on Terror, a combination of pre-existing technologies that was initially deemed useless by the U.S. Department of Defense and the CIA, and only became an accepted implement of war after missions against terrorists were carried out" (Burnam-Fink 2012: 84).

Developed in the early 1990s and first deployed in 1995, the Predator became viable in the post-Cold War period when the United States, now without a "peer competitor," redefined its "threat map" to prioritize "non-state actors." In the Cold War, the Soviet Union served as the principal threat to the United States and NATO and the US surveillance system was organized around monitoring the activities of state actors such as the countries comprising the Warsaw Pact. Satellite surveillance, in this context, became slow and unwieldy, so a more "agile" type of surveillance in the form of spy planes became the new preferred option. Furthermore, the choice of remotely piloted craft meant that concerns of pilot survivability could be addressed more cheaply than would be possible if a conventionally piloted surveillance aircraft were to be designed and built (Burnam-Fink 2012: 86). At the same time, the Predator did not just have to overcome technical issues as part of its concretization. A whole range of social and political problems had to be overcome as well. As an example, Burnam-Fink (2012: 87) notes that Air Force policy regarding the calculation of flight hours needed to be amended to include UAV flight time in order to make UAV operations more attractive to pilots. What Burnam-Fink (2012: 84) effectively demonstrates is that the Predator drone is shaped by social and political forces as well as technical ones and that it cannot be understood in the abstract:

> The Predator drone is more than just a machine; it is the most visible node in a network that binds together pilots at Creech Air Force Base in Nevada,

mechanics in air bases scattered across the globe, soldiers in combat zones, analysts that draw up lists of targets, and operators who decide that an image on the screen corresponds to an intended target. The Predator drone has created new institutions of state power, which formulate missions and in turn demand the continued existence and use of the Predator drone.

What a linear history of the UAV does is take the military drone out of its associated milieu and make it appear to fly out of the past, unconnected to the geo-political present. The popular sweeping linear histories of military UAVs are, in effect, psuedo-historicizations in which the actual material conditions that give rise to and sustain the technical object are obfuscated. A linear history makes the trajectory of the military UAV seem straight and narrow, only impeded by the lack of technical know-how. It does not tell us how our immediate present made the Predator and Reaper drones possible, only that the idea of putting a bomb in a balloon set the wheels of progress in motion so that the use of armed UAVs to extrajudicially kill was something somehow foreseen over one hundred years ago. In contrast, Burnam-Fink's account presents the Predator as something that emerges within a particular milieu, that is an ensemble of social, technical, political, economic, and even cultural forces. The Predator and its milieu are then concretized as a socio-technical object and its associated milieu. The Predator itself is an assembly of technologies that is kept aloft by being part of a larger assembly of heterogeneous elements—what Slack and Wise (2005) term "a network of connections," that brings together materials, flesh-and-blood, institutions, practices, documents, and talk. Burnam-Fink's (2012: 84) description of the drone as the most visible node in a network is extremely helpful because it allows us to begin to think of technological objects not as simply discrete, well-bounded, and finite objects that are meaningful strictly in terms of their defined purpose and, instead, to begin to look at them as an assembly, in which their meaningfulness is (over)determined by the broader set of relations that are brought to bear on the object.

Defining technology as apparatus

Slack and Wise refer to technologies as "networks of connections" to encapsulate the heterogeneity of technologies. They charge that we need to "look past the idea that technology is just the physical stuff" (Slack and Wise 2005: 35) in order to understand how technologies are, in fact, confluences of knowledges, activities, and materials that extend beyond the immediate physical limits of the device. Technical objects are organized, invested with

DEFINING TECHNOLOGY: TECHNOLOGY AS APPARATUS

significance, endowed with values and capacities, and incorporated into social action and relations, and so our discourses on technology are always realized through the interconnection of signification, material artifacts, techniques, and cultural values. What this points to is the need to think of technologies as more than mere things that serve practical purposes. Every technology not only affords accomplishing some task or set of tasks, but it also carries with it a set of expectations about who will use that technology, as well as when, where, and why they might use it. Rather than understanding technologies as being neutral, transpicuous tools, I should like to propose that we, instead, think of technologies as a form of apparatus.

In his essay, "What is a Dispositif?" Deleuze (2006: 339–40) proposes that an apparatus operates along three dimensions. First, there are what he terms the "curves of visibility"; secondly, there are the "curves of utterance"; and, thirdly, there are the "lines of force." Each apparatus is made to be seen and to be spoken about in historically and materially specific ways. The first two dimensions relate to the relationship between knowledge and the apparatus—what can and cannot be visible and spoken to in relation to the apparatus. This means that within every technological apparatus or assemblage there are possibilities but also limits as to how it can be represented. The third relates to the relationship between power and the apparatus, what trajectories or practices are circumscribed through the apparatus. In other words, the technological apparatus organizes an associated range of potential performances or techniques of socio-technical action. In this way, technological apparatuses can be understood as semio-material mediators of knowledge and action. In addition to these dimensions, Deleuze (2006: 340) credits Foucault with discovering "lines of subjectivation," which entails the production of subjectivity in relation to the apparatus. Thus every apparatus is constitutive of subjects specific to the dimensions of the apparatus itself. Hospitals produce patients, mental asylums produce the insane, and prisons produce prisoners.

Agamben declares in "What is an Apparatus?" that "I shall call an apparatus literally anything that has in some way the capacity to capture, orient, determine, intercept, model, control, or secure the gestures, behaviours, opinions, or discourses of living beings" (Agamben 2009: 14). An apparatus creates enduring relations between elements so as to produce a disposition of gestures, behaviors, opinions, and discourses, which is embodied in the technology-subject relation. A technological object never acts alone upon the subject but rather always enters into a relation with the other elements assembled within the apparatus. Agamben cites Foucault's comment that the apparatus can be understood as a formation or network of heterogeneous elements including "discourses, institutions, architectural forms, regulatory decisions, laws, administrative measures, scientific statements, philosophical, moral, and philanthropic propositions" (Agamben 2009: 2). In this way, apparatuses are simultaneously both symbolic-discursive and material. As Deleuze (2006: 342) recapitulates, "Each apparatus is therefore a multiplicity where certain processes in becoming are operative and are distinct from those operating in another apparatus."

The "smart-bra" depicted in Figure 3.7 serves as a very good example of what interconnections can be made visible by such an approach. The smart-bra apparatus is a prototype wearable self-monitoring technology intended to deliver to the wearer just-in-time interventions to prevent emotional eating (Carroll et al. 2013). Equipped with sensors to measure heart rate and electrodermal activity, the smart-bra is designed to monitor the emotional state of the wearer and alert *her* to rising levels of stress. Data are collected by the bra sensors through a smartphone app connected to the bra via Bluetooth and then uploaded to a cloud computing storage system. While this is only a prototype, were it to become a commercially available device, part of the monetization strategy would no doubt include turning the data over for marketing analytics as do other self-monitoring devices. While the bra, as a garment, functions as other bras do, its augmentation with sensor technologies rearticulates it to another set of interconnections of discourses, institutions, regulatory decisions, scientific statements, philosophical, moral, and philanthropic propositions, and so on. One obvious criticism of the device has been the way it reproduces certain assumptions about emotional eating as a gendered dysfunctional behavior and caused by a lack of emotional metacognition. By integrating this technology into the form of a bra, it reproduced those assumptions and made it wearable.[1] As an apparatus of capture, the smart-bra brings together technologies, gestures, behaviors, opinions, and discourses to produce a self-monitoring gendered subjectivity that seeks to help the wearer carefully manage her own emotional states and corresponding eating habits by fitting herself within the smart-bra apparatus. In other words, the smart-bra apparatus is anything but a neutral and transpicuous tool for providing benign support for the mammaries and overeating. Instead, it literally consolidates a constellation of significations, material artifacts, techniques, and cultural values around femininity and eating.

Conclusion

In this chapter, I have sought to lay out a theory of the relationship between discourse, technology, and culture. In order to abandon the dualist view of the technology-culture relation, I have adopted Simondon's theory of individuation. Individuation is very different from individualization. The technical object does not take form or come into being as a discrete and finalized thing as one would expect to happen under a process of individualization. Instead, the process of individuation means that the technical object is an object in becoming. Its form at any given moment is always contingent and provisional. The individuated object is understood to be a moment or event in a lineage that is always in flux. Accordingly, the relationship between technology and culture is preconceived through Simondon's conception of the object-associated milieu relation. This affords us to adopt a non-hylomorphic view of the technology-culture relation and a means to reject the dualist view presented at the start of the chapter.

It was also argued that the technical object should be understood as heterogeneous as well as contingent. Technologies should be understood to be embedded in networks of connections that connect the object itself to a multifarious assembly of materials, discourses, practices, and bodies. Thus, the understanding of technologies as neutral and transpicuous tools was rejected and, instead, the concept of the apparatus was used to describe technologies as a confluence of knowledges, activities, and materials. By rethinking technology as a formation or network of heterogeneous elements that include both symbolic-discursive and material elements, an alternative critical discourse on technology begins to take hold and allows us to critically and comparatively engage with the four discourses on technology that are addressed through the remainder of this book.

2

Multimodal critical discourse analysis

This chapter outlines the theoretical framework for conducting critical analyses of multimodal discourse. Historically, language has been privileged as the principal system of communication and discourse has been largely understood as being a linguistic phenomenon. When texts include semiotics other than language, they have largely been understood as playing a supportive role to the meaning of the verbal text. For example, if a picture accompanies verbal text, it has often been thought to simply illustrate what the verbal text communicates. Logocentrism meant that language was the hegemon among the semiotic systems. This language-first conception of meaning has increasingly been subjected to critical rethinking and considerable scholarship has emerged to challenge this notion that communicators privilege language when interpreting texts that utilize more than one semiotic. It is not uncommon to hear talk of a visual turn in human communication and that images have supplanted language as the new hegemon. But this is, in fact, merely looking at the same coin from the other side. It still presumes that meaning is produced first in one semiotic system and then reinforced by others. Multimodal theories of communication do not presume that one semiotic system always takes center stage and others only play a supporting role. Instead, it is increasingly understood that communication is accomplished through the interaction of semiotic systems or modes and that meaning itself is a recombinant phenomenon.

Accordingly, the aim of this chapter is to detail the theoretical work that has been done in social semiotics to build upon a CDA approach in order to develop MCDA. What I propose to do in this chapter, then, is introduce some of the core principles that underpin CDA and, particularly, the more recent "multimodal turn" in CDS. CDA incorporates a number of different approaches including Norman Fairclough's (2013) Dialectical-Relational Approach, the

Discourse-Historical Approach most readily associated with the work of Ruth Wodak (Reisigi and Wodak 2009), and the Sociocognitive Approach associated with Teun van Dijk (2008), in addition to the Social Semiotic/Multimodal approach, which is foregrounded here. Additionally, CDA is being stimulated by researchers who align its critical practices with cognitive linguistics, such as Christopher Hart (2014), and with corpus linguistics, like Paul Baker (2012). Accordingly, CDS, like CDA, is something of an umbrella term and should not be construed as a narrowly defined method of textual analysis. Instead, as will be discussed in the chapter, it is better to think of CDS as an approach that includes a number of different theories and methods that all share in understanding language (and other systems for meaning making) to be a form of social practice and that communicators resourcefully draw upon those semiotic systems in order to accomplish particular meaning-functions.

The chapter starts by contextualizing the concept of discourse explaining its linguistic and non-linguistic uses. Discourse, it will be argued, functions as resources for constituting what can be known about a particular subject. The chapter then explains the concept of semiotic resource and its centrality in a theory of communication in which meaning is accomplished through the selection and organization of available elements made significant within semiotic systems. Drawing from Halliday's insight that language is one of a number of available social semiotics, the concept of mode is then introduced to highlight how social semiotics such as speech, dance, music, and so on are not only resources for communication but also make available sets of resources for accomplishing representational, interpersonal, and compositional communicative functions. At the same time, it is noted that communicative events rarely entail the use of only one semiotic mode. An act of face-to-face dialogue, for example, typically enlists the modes of speech, gesture, and clothing at the very least. Appropriately, the theory of multimodality is then introduced to address the way in which meaning is produced recombinantly across semiotic modes. Finally, Iedema's (2001: 36; 2005) conceptualization of the process of resemiotization is presented so as to expand the heterogeneity of meaning introduced by multimodality to address how discourse is "translated" beyond texts into what he terms "meaning-materiality complexes."

Discourse: A social semiotic approach

Linguistics has typically understood discourse to be any extended sampling of connected written or spoken language that is "naturally occurring." The primary concern, then, has been to elucidate how linguistic elements or features at the level of the sentence or clause are combined to produce meaning in texts that continue beyond a single sentence. While practitioners of CDA, to varying degrees, work within this tradition (see Wodak 2004), this understanding has also been expanded to encompass Foucault's reconceptualization of discourse as systems of representation that operate beyond sentence grammar (see Hall 2001: 72).

Contrasting his own approach with that of linguistic analysis, Foucault (1976: 30) extends the concept of discourse from a connected series of utterances linked grammatically at sentence level to refer to a grouping of statements or "verbal performances," which are linked by a "body of anonymous, historical rules, always determined in the time and space that have defined a given period, and for a given social, economic, geographical, or linguistic area, the conditions of operation of the enunciative function" (Foucault 1976: 131). The Foucauldian definition of discourse, therefore, differs from the conventional linguistic one by moving beyond a concern with formal rules of combination at the level of the individual text and, instead, looks to the rules of association between texts or statements, which in turn produce "relatively well-bounded areas of social knowledge" (McHoul and Grace 2002: 31). This difference is succinctly described by Gilles Deleuze (2006: 17) in his discussion of Foucault's method:

> He chooses the fundamental words, phrases and propositions not on the basis of structure or the author-subject from whom they emanate but on the basis of the simple function they carry out in a general situation: for example, the rules of internment in an asylum or even a prison; disciplinary rules in the army or at school.

On the surface, then, a genealogical analysis of discourse would seem to address a very different class of rules than those of linguistics-based analyses. This is precisely the charge put forth by Alec McHoul and Wendy Grace (2002: 31) who contend, using Conversation Analysis as their example, that traditional linguistic discourse analysis "looks for techniques of 'saying'" whereas "Foucault's discourse theory looks for techniques of 'what can be said'." However, while this may well be a valid criticism of those more synchronic forms of linguistic discourse analysis, such an assessment of CDA and, more broadly, CDS would only hold true if it were somehow unable to move between the text itself and the broader social structures—in short, if it were power blind.

In point of fact, CDA also approaches texts as communicative performances or events and is equally concerned with "the boundaries of possibility that determine the limits of what can be said, by whom, and in what fashion" (Anaïs 2013: 128). For critical discourse analysts, these boundaries are, in fact, realized in the structuring of textual elements such as lexicogrammar and it is by studying these structures that the analyst is actually able to make claims about a particular discourse. Indeed, as Machin and Mayr (2012: 20) explain, "The process of doing CDA involves looking at choices of words and grammar in texts in order to discover the underlying discourse(s) and ideologies." So while Theo van Leeuwen (2005: 94), for example, directly acknowledges Foucault when he defines discourses as "socially constructed knowledges of some aspect of reality," he also seeks to demonstrate how discourses can be analyzed by a systematic practice of closely and comparatively detailing and describing features of texts in support of broader theoretical claims: "It is on the basis of such similar statements, repeated or paraphrased in different texts and dispersed among these texts in different ways, that we can reconstruct the knowledge which they represent" (van Leeuwen 2005: 95). Accordingly, discourses are immanent in communicative events and do not exist independent of expression (Kress and van Leeuwen 2001).

Evidence for the existence of a given discourse is therefore to be found in its articulation across multiple texts addressing the same subject matter in similar, though not necessarily identical, ways. Returning to Foucault's (1976: 117) exposition of discourse, a given discourse comprises a limited number of statements that can be said to belong to the same discursive formation. In this respect, discourses are understood as mobilizing "certain bits of knowledge" that "are shared between many people, and recur time and time again in a wide range of different types of texts and communicative events" (van Leeuwen 2005: 97). A discourse thus circumscribes how a given subject is to be knowledgeably represented.

So, for example, when *Time* magazine made "You" its 2006 "Person of the Year" as a kind of follow-up to the 1982 "Machine of the Year," the cover announced that "You," the reader, is in control of the pictured personal computer and with it "the Information Age." As an acknowledgment of "the person or persons who most affected the news and our lives, for good or ill, and embodied what was important about the year," (Time 2006) it calls upon the reader to recognize himself or herself (indeed, the computer display is printed with a reflective ink) as the subject in question. As such, the cover constitutes the computer user as one who is in charge at the keyboard and the technology as simply a means of extending human control over our environment. At the same time, we might also want to consider how the cover also limits the user's relationship to computing to being one of active onlooker engaging with what appears on the screen but having little role in determining the conditions under which he or she engages with the computer and the "Web 2.0." In essence, "you" are an end-user and your control is expressed in all of the activities you can elect to do on the Web, but not necessarily in the governance of the Internet itself.

Figure 2.1 Time *2006 Person of the Year magazine cover.*

To be clear, though, discourses cannot simply be equated with representation. Rather, discourses can be said to function as **resources** for representation, allowing communicators to enact socially constructed knowledges in each act of semiosis. In other words, discourses underlie or underwrite the epistemological claims being made within a communicative event. So, from the previous example, the representation of using a computer to access the Internet is underwritten by a conception of the user being like a captain of a ship rather than, say, a worker ant in a colony. It is by drawing from available discourses that we are able to construct "epistemological coherence in texts and other semiotic objects" (Kress 2010: 110). Furthermore, rather than being conceived of as an instantiation of discourse, each communicative event is always understood as both an articulation and a reconstitution of discourse since discourses are both constitutive of social reality and themselves socially constituted. As van Leeuwen (2005: 94) explains, discourses emerge "in specific social contexts, and in ways which are appropriate to the interests of the social actors in these contexts." Discourses, therefore, are more than just ideas or beliefs held about a particular aspect of social reality. Discourses are tied to specific interests and guide how we interpret, act out, and reproduce the social world. Not simply a *product* of the social world, discourses contribute to the reproduction of social relations by constituting the very "situations, objects of knowledge and the social identities of and relationships between people and groups of people" (Wodak 2011: 40).

This means that discourses function to not only inform but also organize and orient social actors. Discourses are, in effect, ontological and axiological as well as epistemological. Discourses not only bring epistemological coherence to representations of social practice, but they also attach particular evaluative ideas and attitudes to those represented social practices. For van Leeuwen (2005: 104–05), those ideas and attitudes can be divided into three kinds: (1) evaluations, (2) purposes, and (3) legitimations. So, for example, were we to watch an "unboxing video" produced by the enthusiastic owner of a new micro desktop computer, we could note how the new computer is evaluated as a consumer electronics device, first, in terms of aesthetic appearance and, second, in terms of technical capabilities, which are in turn incorporated into the aesthetic praise of the device. This, of course, speaks to the purpose of the computing device when it is couched in consumerist discourses of home computing technology where the user acquires the device to "improve" domestic media consumption and bring "technological beauty" into the home. Finally, by framing the talk about the microcomputer in terms of lifestyle enhancement, there is little need to justify the computational overkill of the device since what matters is what it can do for the end-user in terms of aesthetic pleasure and consumer status rather than any particular programmatic tasks.

In this way, the articulation of a discourse serves to both inform and, therefore, legitimate the actions of social actors. As van Leeuwen (2005: 104) puts it, "discourses are never only about what we do, but always also about why we do it." Of course, discourses do not actually determine what can be done or said in an instrumental fashion, but inasmuch as we cannot not represent social reality without drawing upon discourses, they do establish the conditions or frameworks (van Leeuwen 2005: 95) through which the social can come to be known. Furthermore, the relationship between discourse and social practice is such that while discourses legitimate social practices, social practices themselves reconstitute discourses, opening the possibility to discursive and, ultimately, structural change: "Our discourses, our knowledges about the world, ultimately derive from what we do. Our actions give us the tools for understanding the world around us" (van Leeuwen 2005: 102). Therefore, while discourses can serve to legitimate the status quo, they are equally crucial to legitimating social change (see Wodak [2011] and Fairclough et al. [1997]).

It is also important to note that our knowledge of the social and our rationales for why we do the things we do are never accomplished by drawing upon or articulating a singular discourse in isolation. This is why critical discourse analysts tend to pluralize discourse. As van Leeuwen (2005: 94, 95) notes, differently positioned social actors will have diverging ways of interpreting and representing the same phenomenon so as to "include and exclude different things and serve different interests." At the same time, social actors can also draw upon different discourses to address the same phenomenon. It is possible, and, indeed, typical, for the same social actor to use two or more discourses, based in different kinds of authority, when representing a particular aspect of reality. So, recalling the earlier example of the *Time* "Person of the Year" cover, we see, among others, the blending of a discourse of consumer-as-sovereign and one of the computer-user-as-navigator. Both discourses presuppose a subject who is autonomous, rational, and in control but whose agency is limited to using the finished product in prescribed ways and so he or she is always essentially an "end-user." Likewise, returning to the unboxing example, we might note the way the new owner excitedly talks about the device in frames of interconnecting notions of aesthetics, innovation, and convenience. The enthusiastic appraisal of the new device gives expression to these different discourses on technology—different ways of evaluating and engaging with the device—and articulates them within a communicative event. Here the video-maker is calling primarily upon an authority largely based upon lifestyle expertise, but also to some degree on technical expertise. In this way, it makes little sense to reductively refer to *the discourse* of some phenomenon such as home computing or consumer technology and far more sense to ground our descriptions of discourses in their own specific contextualizations and to refer, instead, to the *discourses* (plural) that are drawn upon as frameworks to represent the phenomenon.

Semiotic resourcefulness in communication

While discourses function as resources to be drawn upon by communicators, allowing them to meaningfully represent and interpret some aspect(s) of reality, the way in which discourses are actually realized or semiotized in communicative events is accomplished through the selection and combination of semiotic resources. Semiotic resources refer to those signifying elements or options that are available to us in order to compose and interpret texts. The concept of semiotic resource is preferred to that of the "sign," as it is used in mainstream semiotics, because it affords greater consideration of the diachronic or, perhaps better, resourceful features of signifying practices.

The sign relation of signifier and signified as de Saussure (1959) theorized it, is dyadic and tends to encourage a notion of a relatively static system in which signs have pre-assigned meanings. So, for example, the signifier, *arbour*, in the French language, signifies the concept of *tree*. This relationship is theorized by de Saussure as purely arbitrary and, therefore, entirely a product of tradition. The durability of the sign relation and, ultimately, the system is ensured because "the arbitrary nature of the sign is really what protects language from any attempt to modify it" (Saussure 1959: 73). Since the meanings, once adopted, are, in effect, determined by the system or code, the Saussurean conception of semiotics is said to be synchronic or to privilege the system of the language over its actual use. Semiotic resources, in contrast, are understood to have meaning potentials that are then determined *in use* and *in relation* to the other resources selected by the communicators.

Rather than a dyadic view of meaning, van Leeuwen (2005: 4) describes semiotic resources as "signifiers, observable actions, and objects that have been drawn into the domain of social communication and that have a *theoretical* semiotic potential constituted by those past uses that are known and considered relevant by the users on the basis of their specific needs and interests." Like signs, semiotic resources do have associative meaning but that meaning is more potential than referential. This means that resources have a range of potential meanings in contrast to a clearly defined referential meaning and the meaning potential of semiotic resources rests, then, both in convention and in what current circumstances can afford. The agenda here is to move beyond a focus on codes and signs to stress, instead, the way people draw from available semiotic resources to both produce and interpret communicative artifacts and events. So while signs are typically construed as meaningful because they have an already attributed meaning, semiotic resources have, instead, meaning potential, which suggests that the meanings of resources are never finalized but always *in-process* as part of an open, dynamic reserve rather than a closed, static system.

In contrast to the arbitrary yet fixed conception of the sign, semiotic resources are, instead, conceived of in terms of the kinds of meanings they can afford the communicators. The concept of affordance is borrowed by van Leeuwen (2005: 4–5) and Kress (2009: 55, 58) from James Gibson (1979) and his influential work on perception and attention. For Gibson (1979: 127), "the affordances of the environment are what it offers the animal, what it provides or furnishes, either for good or ill." Affordances are very much like resources, then. They are not simply present in the environment, they are available to the animal. Gibson (1979: 127) nominalizes the verb, afford, in order to convey "the complementarity of the animal and the environment." In other words, an affordance instantiates the relationship between the animal and its environment. The affordance is never purely of the environment itself since it is also implicated in the animal's presence in that environment. Gibson gives the example of a seat that can take a number of different forms such as a chair, couch, or a bench. What matters is that it has the properties that would make it "sit-on-able." Some of these properties are objective such as a flat, firm, and horizontal surface, but others are subjective such as being knee-level relative to the person looking to sit down. As Gibson notes, to an adult, a chair designed for a child will not afford comfortable sitting in the same way as one designed for an adult. The degree to which the object is a resource for sitting then is determined, in part, in relation to the perceiver and not just by the properties intrinsic to the object.

Van Leeuwen compares this concept to that of Halliday's theory of meaning potential. As van Leeuwen (2005: 5) explains, in the Hallidayan conception, "linguistic signifiers—words and sentences—have signifying *potential* rather than specific meanings, and need to be studied in the social context. The difference is that the term 'meaning potential' focuses on meanings that have already been introduced into society, whether explicitly recognized or not, whereas 'affordance' also brings in meanings that have not yet been recognized, that lie, as it were, latent in the object, waiting to be discovered." In this way, by drawing upon the concept of affordance, van Leeuwen is able to further refine our understanding of the meaning potential of semiotic resources. The meaningfulness of any given semiotic resource is always a product of a relation between objective and subjective perceptions of the resource, between communicators and their environment or context. So while for Gibson (1979: 129) an affordance "is equally a fact of the environment and a fact of behavior," we can argue likewise that the affordance of a semiotic resource is equally a fact of the semiotic environment and a fact of the user. Unlike signs, resources have no fixed meaning but, equally important, their meaning is not simply a matter of a pre-established range either. Instead, while semiotic resources will have historically attributed meanings, those already established meanings cannot exhaust the potential meanings that a given semiotic resource might further afford. In the last instance, semiotic affordance is always a property the interaction between context, resource, and user.

The ways in which a semiotic resource affords meaning in terms of both how it has been used in the past and how it can be used is well illustrated by the misattribution of a US fraternity greeting sign as a gang sign. In the summer of 2014, Darren Wilson, a white Ferguson, Missouri, police officer, shot and killed Michael Brown, an unarmed African-American teenager. After the local police were accused of grossly mishandling and using excessive military-style force against protesters, the Missouri Highway Patrol was assigned to police Ferguson with Captain Ron Johnson in charge. Two photos of Johnson posing for photographs with two different African-American men (neither of them dressed in any gang-related attire) were posted on CNNs iReport, a news crowdsourcing platform (read free-labor news gathering), accusing Johnson of being a gang member (McDonald 2014). In the photos, the men made a hand gesture by using the index finger and thumb to form a circle and leaving the other three fingers extended outward. The images then circulated widely over the social media platform Twitter with the same accusations appended. It was soon revealed that Johnson was, in fact, making a greeting sign for Kappa Alpha Psi, a historically black fraternity of which he and the other men were members. Clearly, the gesture afforded the meaning of greeting and membership to those who were members of the fraternity or familiar with the fraternity rituals in general. Ironically, a photograph of Michael Brown making a peace sign circulated after his death had solicited similar responses, by those who saw the gesture as an obvious gang sign. While not all viewers unfamiliar with the fraternity sign interpreted Johnson's gesture as a gang sign, and for many it may not have even been particularly salient, there was clearly a significant group of people who were intensely interested in the hand gesture and for them it afforded a very different meaning. Drawing upon a racist discourse that characterizes black males as innately dangerous and prone to criminal gang behavior, they were unable to see a black man, regardless of his status, making a hand gesture for any other reason than affirmation of gang membership. The hand gesture was obviously salient but because it was made by the hand of a black male, the accusers could only interpret the gestures, seemingly any form of hand gesture made by either Johnson or Brown, as gang signs. In terms of the concept of affordance, what this illustrates is precisely the ways in which a resource such as a hand gesture can have different meaning potentials across different constituencies.

At the same time, as van Leeuwen (2005: 5) reminds us, this does not mean semiotic resources can be freely used to mean whatever the communicators might please. Semiotic resources possess a degree of durability in that their meaning potentials "are socially made and therefore carry the discernible regularities of social occasions, events and hence a certain stability" but, at the same time, mutability as well, since "they are never fixed, let alone rigidly fixed" (Kress 2010: 8). This *metastable* quality of semiotic resources is made clear by Kress (2010: 8) when he writes: "*Resources* are constantly remade; never wilfully, arbitrarily, anarchically but precisely, in line with what I need, in response to some demand, some 'prompt' now—whether in conversation, in writing, in silent engagement with some framed aspect of the world, or inner debate." Semiotic resources, accordingly, are available to us because they have social significance; their significance lies in their being made *potentially* meaningful in historically particular ways but this meaningfulness is never (pre)determined or static in the way that is attributed to signs as part of a code. By privileging the concept of resource over intrinsic systematicity or code, social semiotics is able to attend to the ways in which social actors improvise as well as regulate the use of semiotic resources.

Consider the use of the silhouette image that appeared in the iconic iPod advertisements from 2003 to 2008. The initial television commercial, *Beat* (2001), introduced the iPod in North America depicting a man loading and then dancing to his iPod while leaving his apartment. In the words of Ken Segall (Segall 2012: 92), then Ad Agency Director for Apple, "it was somewhat uncomfortable to watch, and on the web some started referring to it as the 'iClod' commercial." The subsequent ads devised by the agency, TBWA\Chiat\Day resolved this, in part, by using silhouettes rather than real people. However, rather than assuming that the silhouettes in the ads refer to some predetermined signifier independent of use, the notion of semiotic resource supposes that the meaning potential of the silhouettes is derived from both previous conventionalized uses and those possible uses afforded to users within the specific social context in which the communication takes place. In the case of the Apple "Silhouette" ads, the selection of a silhouette over a real person dancing overcomes the difficulty of using a specific individual to serve as an everyperson. While on the one hand, the actor in the *Beat* ad appears to be and dances like just a "regular Joe" with whom we can readily identify, on the other, the "realness" of the actor also potentially undermines the attribution of the "cool factor" being to the product by putting our own cloddishness in our faces. The silhouettes, by contrast, allow actors with not so everyday dancing talents to serve as proxies for the average potential iPod owner. In this way, we can say that the silhouette functions as one of the available semiotic resources that are drawn upon in the composition of the iPod ads.

In adopting the silhouette, what the ads do is draw upon the semiotic potential of a longstanding and well established visual convention of treating the human shadow as a metaphorical "index" for the inner qualities or essence of the individual. Nancy Forgione (1999) has argued that in Paris, during the 1880s to 1890s period, this metaphoric convention gained new significance within aesthetic discourse as the shadow began to function as both an abstract concept and a concrete device such that it "provided a technique for visualizing that helped dislodge old habits of seeing that relied too much on superficial appearances; moreover, it was believed to facilitate access to the intrinsic nature of things" (Forgione 1999: 492). Forgione

Figure 2.2 *Apple's 2001* Beat *commercial.*

speculates that this new significance might in part have been due to the increased nighttime illumination that came with the change-over from gas to electric lights in Paris during that period. In drawing upon this convention, using the backlit profile, rather than the fully exposed frontlit figure to stand for the mobile and, therefore, liberated music consumer, Apple and its agency opted to utilize the "traditional equation of shadow with the inner self rather than the external descriptive appearance of an individual" so as to realize the "Mallarméan ambition of many late-nineteenth-century artists "to suggest" rather than to describe their subjects" (Forgione 1999, 491–92). Similarly, the revised silhouette ads do just this, they suggest rather than describe the iPod user and at the same time cut through appearances to reveal the user in her or his "essence" as animated by music. However, what this essence is, is not solely signified or determined by the silhouette itself.

Convention, therefore, allows the silhouette to be selected and used as a convenient visual metaphor for suggesting the vital qualities of the figure being profiled. The full meaning of the metaphor, however, does not rest solely with convention but rather is fully realized in the ways in which the silhouette form is incorporated into the actual advertisements. Thus, it is the silhouette in combination with the other selected resources that cues how we are to "read" the silhouetted iPod users. The iPod users engage in bold and lively dance moves in empty space, they wear youthful and hip "street fashions," they are backgrounded by equally bold and vibrant colors, the music is invariably quick in tempo, the verbal text is minimal, and the typeface is a clean, bold, white sans serif, and of course, there is the placement of the contrasting white iPod and ear buds. And, while some ads did include headliner bands and musicians (more monochromatic than in silhouette), the dancers/iPod users themselves always perform alone. Taken together, it is the combination of these resources that helps to define or overdetermine the content or the "essence" of the silhouette figure. It comes to suggest someone who is vibrant, independent, and dancing to the beat of his or her own music player, and, most importantly, it "figuratively" leaves a space in the image where the viewer can imagine herself or himself. These elements function, therefore, as semiotic resources that are selected and used in accordance with the potential meanings that they afford.

However, this example also illustrates how the concept of semiotic resource is important in a second crucial way: not only does it highlight a more dynamic and open process of meaning making, but it also invites us to recognize that meaning making and, ultimately, discourse are not reducible to language. The meaningfulness of the advertisements actually depends very little upon language inasmuch as the wordings themselves—the lexicogrammar—functions primarily to anchor (Barthes 1977) the iPod+iTunes product and the Apple brand within the experience of the ad. Instead, it is the resources derived from the other semiotic modes—moving image, videography, gesture, choreography, attire, and music—that for the most part constitute the meaning of the ad for the viewer. The silhouette ads are therefore a good example of what can be termed cross-semiotic or multimodal communication whereby multiple semiotic modes are enlisted in the composition of a text.

Mode and metafunction

What exactly counts as a semiotic mode and how mode should be defined has proved somewhat contentious (see Jewitt [2009a: 21–22], Machin [2013: 3], and Forceville [2013: 51]); however, Kress (2009, 2010) provides a functionalist account of mode that will serve our purposes here. Modes, then, will be understood as a special category of "socially shaped and culturally given" (Kress 2009: 54) resources that social actors regularly draw upon for communication. At the risk of putting it blithely, a semiotic mode is whatever can function as a semiotic mode for a given social group in order to meet their communication needs. Kress (2009: 54) acknowledges that while all cultural forms and practices "bear" meaning, for example, objects of material culture, modes are limited to those phenomena that have communication as their principal function. Obvious examples of semiotic modes would include speech, writing, image, moving image, music, gesture, and so forth. These are means of communicating, each with its own potential for meaning making, that have come to be used in regularized ways within specific sociocultural groups.

Semiotic modes are, therefore, both material and symbolic since their meaning potential is always determined in relation to their material affordance. This is made clear by Kress and van Leeuwen (2001: 28) when they note that "in semiosis, the materiality of modes interacts with the materiality of specific senses, even though modes are conventionalizations produced through cultural action over time, and therefore abstract in relation to any one particular action." In this way, while a mode may privilege one sensory perception such as sight, hearing, touch, and so on, modes cannot be conflated with the senses. And while it might also seem appropriate to think of modes as "channels," such an understanding erroneously depends upon thinking of modes as conduits through which to pipe messages as might be presupposed in a transmission model of communication. Finally, modes are not reducible to media either, since a single medium such as print or film will often accommodate more than one semiotic mode (Kress and Van Leeuwen 2001: 6, 22). Instead, we should think of semiotic modes as meta-resources, since each mode both entails a set of associated resources for meaning making and is a resource in its own right. Accordingly, modes do not simply allow addressors to pass meaning through to addressees. Rather, modes make available resources that are specific to the materiality of that mode and so enable, but also constrain, the potential meanings that can be made by communicators. In this way, imagining the mode and the message itself to be distinct fails to consider how texts are composed through the material and semiotic affordances of the mode.

Kress (2009: 56) uses the term *modal affordance* to refer to what a particular mode may be capable of expressing in communication. Modes, like all semiotic resources, bring different meaning potentials based upon both the ways in which they have been enrolled into communication practices (historic use) and also the materiality of the modes themselves (affordance). Different modes will have distinct affordances and thus will differently "enable specific semiotic work drawing on these affordances" (Kress 2009: 56). What this means is that modes are not simply available to communicators to make meanings in the same way or to "carry" the same meanings across modes. Since meaning is realized through drawing upon the resources that a given mode makes available, meaning is always both accomplished through and limited by the articulation of modal resources. For example, writing employs words, syntax, clauses, sentences, and so forth, which are organized into grammars, while images draw upon visual elements that are organized into compositions. Writing affords the expression of argumentation very well while an image, though it can be persuasive, does not lend itself to the same form of expression. Semiotic modes, accordingly, are never neutral carriers of meaning since each mode is historically and materially constrained in the kinds of meaning potential it can afford communicators. Modes are selected and drawn upon in accordance with their potential to accomplish necessary or needed meaning-work (Kress 2009: 57). Modal affordance is thus a part of the conditions under which communication takes place and so shapes not only how meanings are expressed but also what meanings can be expressed.

As semiotic resources, modes are utilized in order to accomplish three fundamental communication tasks: (1) produce representations, (2) constitute interactions between social actors, and (3) compose those representations and interactions as specific kinds of texts. In other words, each act of semiosis or communicative event is simultaneously a representation, an interaction, and a composition, and communicators must perform these tasks by selectively drawing upon the resources that modes make available. Halliday (1985: 53) proposes that these tasks are realized semiotically through three metafunctions: the ideational, the interpersonal, and the textual. While Halliday has primarily focused upon how these metafunctions are accomplished in clausal structures, Kress and van Leeuwen (2006) (see also van Leeuwen 2005) have, in turn, extended their application to other semiotics beyond language such as images and music and renamed them as the representational, interactional, and compositional. Accordingly, all semiotic modes have the potential to realize these three metafunctions, though not necessarily in identical ways. Bezemer and Jewitt (2009: 7) have referred to this intrinsic functionality of semiotic modes as the "meta-function test." The important thing to remember though is that the ability to meet the conditions of the test is not so much dependent upon qualities inherent to the mode itself but rather that there is a community of users who are able to draw fully upon the mode in order to accomplish the three communication tasks.

Modes function representationally by constituting what is taking place through the depiction of social action and actors. Attending to language, Halliday (1985: 101) demonstrates how clauses can be analyzed in terms of transitivity, which is a basic semantic framework comprising (1) the process itself, (2) participants in the process, and (3) circumstances. Again, this way of conceptualizing representation as function can be extended from written and spoken language to other modes, allowing us to compare how modal affordances are used in the creation of transitivity structures. As van Leeuwen (2008) demonstrates, transitivity structures semiotize represented actions or activities as types of processes, identify (or obscure) actors, and allocate agency among represented actors. Different modes will have associated resources for semiotizing transitivity elements such as agency or actor. In Indo-European languages, the actor-process-goal structure tends to attribute agency to the represented participant that occupies the antecedent "actor" position. For example, in the clause, "the man vacuumed the carpet in front of the woman," "the man" occupies the role of actor within the clause and is represented as the agent responsible for vacuuming. The woman functions as a represented participant but consigned to the circumstance under which the man does the vacuuming, her agency is somewhat diminished. Images, by contrast, do not use the same sort of syntagmatic forms for representation. For Kress (2009: 56), linguistic and visual modes make available differing semiotic resources such that "where writing or speech has words, image has 'depictions.' It uses basic visual entities—circles, squares, lines, triangles. Words are spoken or written; images are 'displayed'; the 'site of display' is a semiotic entity characteristic of image(-like) communication." So if we, instead, look to Figure 2.3 in which we see an image of a man vacuuming a carpet while a woman reclines watching, we can consider how it depicts the same process through a different spatial organization of its elements.

In *Reading Images*, Kress and van Leeuwen (2006: 66) borrow from Arnheim to detail how narrative visual images, which represent some sort of social action taking place, can be understood in terms of volumes and vectors. Participants are conceived of as volumes that are understood as discrete entities, "which are salient ('heavy') to different degrees because of their different sizes, shapes colours and so on" (Kress and Van Leeuwen 2006: 49), but also in relation to one another because they are in effect connected through vectors that visually realize the processes that the participants/volumes undergo. In the vacuum ad, there are two primary volumes: the vacuuming man being one and the reclining woman being the other. The man faces the woman in the center right of the photographic image and is the much larger and heavier volume, while the woman is further back in the image and is smaller. However, there is no precise and clear meaning that can be derived from the photograph in the same way that we find in the clause "the man vacuumed the carpet in front of the woman." As Machin (2011: 169) points out, that specific meaning "only *appears* to be there when we produce verbalizations of images that allow this to be done." Someone looking at the ad image could quite easily verbalize, "The man vacuumed the carpet while the woman watched," or "The man chatted to the woman while he vacuumed," or even "The woman put down her book and watched as the man vacuumed the carpet." Clearly, this difference in representational meaning potentials is a matter of modal affordance and semiotic modes do not, therefore, simply serve as channels for delivering the same meanings. At the same time, this does not mean images can mean whatever an individual viewer might will it to mean, since meaning is always ultimately a social activity. Rather, it demonstrates that images afford meanings differently from language. Given the gendered distribution of unpaid domestic labor means that women in a heterosexual relationship are more likely to be the ones doing the actual vacuuming; the significance of the way the act or process of vacuuming is being represented here will be interpreted quite differently than in a society where domestic labor is more equitably distributed.

Figure 2.3 *Magazine advertisement for vacuum.*

The second communication task that must be accomplished through modal affordance is the realization of social relations, or what Halliday terms the interpersonal metafunction. Every communicative event is not only an act of representation but also one of exchange. In other words, modes make available semiotic resources that are selected in order to semiotize social relations as interactional meanings. As Kress and van Leeuwen (2006: 48) note, every communicative event entails two types of participants: the represented participants within the text and the interactive participants actually engaged in the act of communication. Texts, therefore, potentially semiotize three kinds of interactions or social relationships: first, between the interactive participants; second, between the represented participants when more than one are present; and, third, between the represented participants and the interactive participants. So in the case of the declarative statement, "The man vacuumed the carpet in front of the woman," there is an impersonal exchange of information between interactants that is presented as an offering of a given fact, without any particular attitudinal orientation being implied, that can be either accepted or rejected. On the other hand, were the statement to, instead, read, "The husband vacuumed the carpet for his wife," the circumstance has changed such that not only are we given to understand that the two participants are in a married relationship, but we might also assume that the man is doing a task as a favor to *his* wife and that the task is normally hers. A statement such as this one can tacitly affirm traditional gendered relations without coming out directly and communicating that normally women should be responsible for operating the vacuum and cleaning the home.

In images, interactional relationships between viewers and represented participants are realized through three key resources: gaze, social distance, and angles of interaction (see Kress and Van Leeuwen 2006; Machin 2011). Gaze refers to the kind of interaction between represented and interactive participants that is realized in the image. Depending upon whether or not the represented participant(s) return the gaze of the viewer, the image is said to either demand (gaze) or offer (gaze absent). Kress and van Leeuwen (2006: 117–18) understand these as image acts that are akin, but not equivalent, to speech acts in spoken and written language. While spoken and written language can, according to Halliday (1985), afford four basic speech functions or acts (demand information, demand goods and services, offer information, and offer goods and services), images, in turn, afford just the two image acts (demand and gaze). Social distance suggests that there is a metaphorical relationship between image-subject distance and personal space (Hall 1969) such that the close-up suggests intimacy while the long shot, in turn, is suggestive of impersonal social distance. Finally, point of view realizes relations of involvement and power and is largely accomplished through the angle of perspective. Point of view is accomplished through two key axes: horizontal angle, which realizes the degree of involvement, and vertical angle, which realizes power relations. The degree to which the represented participant faces the viewer or not determines involvement. A low-angle perspective tends to elevate the represented participant in status whereas a high-angle perspective tends to diminish the participant's status vis-à-vis the viewer. Importantly, Kress and van Leeuwen (2006: 148) remind us that these features are part of a simultaneous system and so function together to produce the interpersonal meanings of the text.

Returning to the image of the Miele® vacuum advertisement, the man is standing in the foreground with his body turned back toward the woman sitting further back on a couch. The woman is sitting face-forward but is looking in the direction of the man. As an image act, there is no returned gaze between the represented participants and the viewer, so this can be characterized as an offer image. While the man occupies what can be considered a social distance (not intimate, but not impersonal either), the woman sits at an impersonal social distance, which further redounds with the impression that she interacts with the man and not the viewer. From the viewer's perspective, the represented participants are positioned at an oblique angle and at an upward vertical angle. The oblique viewer angle depicts the man and woman as being directly engaged with each other while the viewer is positioned more as a detached onlooker and so realizes a low degree of viewer involvement with the participants. The low vertical viewer angle also gives the effect of the onlooker looking up at the couple, which can suggest the couple be viewed as elevated in status as might be the case were the couple to represent a lifestyle the viewer is expected to aspire toward. Since the man is standing and the woman is sitting, her positioning is less pronounced on the vertical angle than the man's. So while on the one hand, the man is placed in a less traditional role of doing housework (while his partner rests), that "challenge" to convention is actually mediated by the man's stature vis-à-vis both the viewer and the woman.

Finally, modes function to materialize the representational and interactional meanings as "message entities—texts—coherent internally and with their environment" (Kress 2009: 59). One resource for accomplishing compositional meaning visually is salience. Returning to Arnheim's conception of volumes, salience describes how different elements within a composition are balanced, but also ranked in terms of significance or importance. Rather than presuming each volume or visual element to be a distinct entity, each element is understood to bear its own relative "weight" in relation to the other elements of the composition. Salience is derived through the use of resources such as size, placement, depth of focus, color, and so on, which work to either enhance or diminish the visual significance of each element. Looking to the photographic image of the man and woman in the living room as a composition, the positioning of the standing man in the foreground makes him much larger relative to the sitting woman. Note also that the man is wearing dark colors that contrast with the color palette of the room while the woman wears light colors so that her top blends with the couch and wall and her pants blend with the cushion. At the same time, she occupies the center of the photograph and her face is exposed whereas the man's is not. This can have the effect of mitigating somewhat the salience of the male figure. The vacuum being promoted is obviously important to the composition but typically will be less salient than the two human participants. Generally, human faces and figures have a higher salience than non-human objects (Sharma et al. 2009). The vacuum is positioned so that it sits below the woman and rather than being highlighted as the central and most important element in the promotional photograph, its centered position makes it more of a bridge or mediator between the man and the woman. The colors stand out from the white carpet (the yellow is echoed in the two awards in the bottom left of the ad) but do not command the same attention as the man's darker clothing. In effect, like much domestic technology, it disappears into the background of everyday life. Compositionally, the organization of salience in the image works to highlight the human participants and their relationship while suggesting all the while that it is the vacuum itself that underwrites it all.

Multimodal discourse and the process of resemiotization

While the legacy of Parisian semiotics has been to treat the various semiotic modes—image, sound, writing, etc.—as distinct forms that "bear" meaning, the recent emphasis upon multimodality in CDA has highlighted the ways in which texts are often made meaningful through the interaction of multiple rather than single semiotic modes. Thus, a multimodal approach to representation does not simply assume the relationship between modes to be one of redundancy whereby each mode simply replicates the meaning constituted in the other modes. This is not to claim that modal redundancy does not occur but rather that in multimodal texts, meaning is understood as an emergent property of the combination of modes and the collocation of their selected resources. In this way, a multimodal approach to discourse analysis tends to compare and contrast semiotic modes and explore the forms of cohesion that are available to both produce and interpret texts. Thus, in the iPod Silhouette commercials, the attributions that viewers were expected to endow the dancing silhouette with were to be derived from attending to all of the incorporated semiotic modes, or what is more oft referred to as multimodality.

As Kress (2009: 54) notes, multimodality is not a theory of communication so much as it "maps a domain of inquiry." This emphasis upon exploring meaning-making practices across semiotics is sometimes characterized as "the multimodal turn," which came to the fore in the late 1980s and continued into the 1990s with social semiotic scholarship that sought to combine the Hallidayan-inspired systemic-relational approach to semiosis with the analysis of texts that draw upon semiotics other than language (Iedema 2007: 32).

The impetus behind this turn, according to Iedema (2003: 33), "centres around two issues: first, the de-centring of language as favoured meaning making, and second, the re-visiting and blurring of the traditional boundaries between roles allocated to language, image, page layout, document design, and so on." So while initial foundational work such as Kress and van Leeuwen's (2006) *Reading Images: The grammar of visual design* faced some criticism as being an over-extension of linguistic analysis to non-linguistic modes of communication (see Machin 2011), as it has evolved, multimodal discourse analysis has by and large moved beyond language-centered analyses while retaining a systemic-relational approach. And while a multimodal approach to discourse analysis "rejects the traditional almost habitual conjunction of language and communication" (Jewitt 2009b: 2), it does not diminish the constitutive role played by language as a social semiotic in communication and the reproduction of social relations. As Jewitt (2009b: 2) continues, "A key aspect of multimodality is indeed the analysis of language, but language as it is nested and embedded within a wider semiotic frame." Even though language is a key aspect of multimodality, it is not intrinsically the principal one. This is why Machin (2013: 5) cautions that in multimodal discourse analysis, we should be careful *not* to imagine that the meaningfulness of texts is determined through a "language-first kind of sequence."

What this means is that language is, instead, one of many resources that communicators draw upon as available and, in multimodal texts, the relationship between modes is frequently heterarchical rather than hierarchical. By heterachical, I mean that in multimodal texts, the individual semiotic modes always have the potential to be ranked in a number of different ways. While certainly in some cases, language can function to fix or clarify the meaning of a multimodal text—what Barthes (1977: 38–41) terms anchorage—in many texts, the verbal component of a text cannot be attributed with such a determining or overriding function. So although Barthes proposed that the relationship between image and text can be characterized as either anchorage or, less commonly, relay (verbal text is ancillary or complementary image), current research in multimodality tends to eschew such straightforward binarisms. In other words, while language might have been historically privileged as the primary mode of communication (and discourse), outside of mono-modal texts, this primacy is much less certain.

This recognition that our communications are enacted in a much wider semiotic frame means that it is erroneous to privilege discourse as a linguistic phenomenon, which is then simply re-expressed in other modes. This is clearly exemplified by van Leeuwen's (2004, 2005: 120–22) analysis of the Lord Kitchener recruitment poster where he demonstrates how the different semiotic modes of clothing, gesture, image, language, and typography are articulated as a "hybrid" communicative act, which is more than what is communicated by the verbal text. Since discourse is not reducible to any one mode, such as language, the realization or semiotization of discourse is accomplished through the overdetermination of meaning inter-semiotically or multimodally rather than simply through linguistic determination. It is for this reason that discourse should not be thought of as a "language-first" phenomenon. The meaningfulness of multimodal texts is, therefore, accomplished through recombinant relations between semiotic resources across modes rather than referential relations between signs operating in distinct semiotic modes. In this way, discourse as a concept is reconstituted from "language-based cognition . . . to multi-semiotic social practice" (Iedema 2007: 933). However, not only does the concept of multimodality offer a means to engage in a systemic-functional approach to meaning-making practices without sliding back into a logocentric conception of semiosis, but it also points to the intrinsic heterogeneity of discourse.

Since discourses are always invested in texts, practices, and objects, they are therefore intrinsically material and, therefore, heterogeneous. While multimodality has emphasized the heterogeneity of meaning potentials at the level of the text, there also needs to be a means of addressing how discourses are communicated beyond the level of the text. Drawing in part from Jackobson's conception of intersemioticity and Actor-Network Theory concepts of delegation and translation, Iedema has introduced the term of resemiotization to describe the process by which meanings are translated from one domain to another into increasingly more durable and, therefore, presumably less negotiable materialities. Resemiotization highlights "the role of material reality in communication" (Iedema 2001: 24) and the ways in which, through delegation, the material is expressive and the expressive is, in turn, materialized. Looking at communication through the lens of resemiotization necessitates explicitly attending to "how meaning making shifts from context to context, from practice to practice, or from one stage of a practice to the next" (Iedema 2003: 41) as a series of delegations or displacements, and the practices of meaning making are, therefore, intrinsically intersemiotic, or what is more frequently termed multimodal. But these delegations are not limited solely to what have been typically thought of as semiotic modes (talk, writing, images, music, etc.) and, instead, extend to the material itself. As an example, Iedema cites Bruno Latour's (1992: 235) example of the door closer and the delegation of the program of keeping the door closed from verbal request, to written notice, to the installation of a hydraulic door closer. Thus, in drawing from Actor-Network Theory, Iedema (2001: 36, 2005) foregrounds the intrinsic heterogeneity and materiality of semiosis by characterizing it as being generative of "meaning-materiality complexes."

Resemiotization entails a process of translation from one domain of activity to another that results in two things happening. First, "issues of localized difference and concern" become abstracted as they are translated "into specialized and technical discourses and practices" (Iedema 2001: 24). Second, these abstracted discourses and practices become increasingly durable and resilient as they are inscribed not only in texts but also in other materialities such as objects and practices (Iedema 2001: 24). For example, in the example of wearable fitness trackers, these devices can be understood as translating wearer movement into data that is then reconstituted as, in the case of Nike, "Fuel." Individuals seeking to self-monitor their physical activity are given feedback in the form of a score that abstracts life into digital matter (Thacker 2004). "Fuel," for example, is not so much a measure of energy expended as it is of points accumulated. Additionally, points earned by meeting prescribed score levels are used in awarding virtual trophies. Through these self-monitoring systems, the movements of the individual body throughout the day become a numerical value, which then can be used to self-assess one's performance as "healthy" subject. Accordingly, these devices make durable the belief that health is a private matter of individual responsibility and that through "healthy" competition, health and wellness are analogous to a marketplace. In keeping with Delueze's (1992) notion of the dividual derived from the disciplined individual, the actions of the individual body are not only trackable but can also be separated out and quantified into data that can then be recombined in any number of ways. Furthermore, this discourse of health as individually earned or squandered is inscribed not only in the app interface but also in the personal device and settings in which it can be used. In this way, the wearable fitness regime can be understood as a meaning-materiality complex, which realizes a commodified version of health across a network of texts, practices, and objects. Therefore, when we attend to the process of resemiotization, we are, in effect, conceptualizing semiosis as an emergent, heterogeneous, relational, and recombinant phenomenon.

Conclusion

This chapter has proposed an MCDA approach that is derived from a broad range of scholarship that can be grouped together as CDS. CDS encompasses CDA, social semiotics, and Foucauldian Discourse Analysis, among others. All of these approaches share in the adoption of Foucault's reconceptualization of discourse whereby it is understood to be both constituted by and constitutive of the social. They also share in a concern for the role discourse plays in the reproduction of asymmetrical power relations and the various forms of subjugation and exploitation that are legitimated and reproduced through discourse. An MCDA approach retains these commitments and extends the research into the role that discourse plays in semiotizing the social and legitimating the actions of social actors beyond language to include non-linguistic phenomena.

Multimodality entails looking at meaning making as a social practice in which meanings are realized through the articulation of multiple semiotic modes. Communication is rarely limited to a single mode and multimodal discourse analysis seeks to explore how meaning is realized through the interaction of modes as well as the selection and organization of semiotic resources within a specific mode. As such, the relationship between modes is not necessarily hierarichical and can, in fact, be heterachical. Different modes have different materialities and affordances and so multimodal texts are best understood as meaning-materiality complexes. In this way, semiosis should be conceptualized as an emergent, heterogeneous, relational, and recombinant phenomenon.

3

Analyzing multimodal discourse: A toolkit approach

Practitioners of CDS tend to make explicit political and epistemological claims regarding their approach. CDS and, more explicitly, CDA are understood as being directly linked to Critical Theory and the Frankfurt School and, therefore, part of a broader emancipatory project. Through critique as the dialectical application of theory and method, CDS is implicated in not merely describing meaning-making practices but also intervening in them so as to foster real social change. This requires a form of critical practice of reading texts against the grain. To do this, we first need to consider what we mean by being critical and the expected political outcomes of CDS as praxis.

To the extent that CDS is understood as demystifying the semiotization of social relations, the problem of voluntarism comes to the fore. To what extent does CDS privilege itself as "knowing better" than those it considers to be allies? If the critical project of CDS is framed as one of demystifying power relations that otherwise remain hidden to social actors raises some ethical-epistemological and, perhaps, even ontological problems for those doing the work of CDS. This, I believe and intend to argue in this chapter, is easily offset if we return to Foucault's concepts of positivity and visibility. Furthermore, I propose to argue that positivity is not in any way incommensurable with CDS' critical orientation.

In the ensuing chapters, I will be detailing some of the key discourses that have come to shape how we commonly tell stories about technology and its consequences for our everyday lives. This chapter lays out the methodological approach that will be adopted in order to develop a critical study of technological discourses and their articulation within our everyday technocultures. It begins by addressing the matter of being critical within CDS practice and situates it in relation to Latour's notions of "matters of fact" and "matters of concern." In addition to the Frankfurt School, CDS draws directly from Foucault's own theorization of discourse and this is presented as a solution to the problem of tacitly inheriting a kind of camera-obscura epistemology in which ideology is understood as the world turned upside-down and a direct expression of those in a position of privilege. The chapter then rejects the supposed contrariety between description and critique, arguing that description can be enlisted as a tool of critique when the two are applied sequentially. This leads to adopting Machin's proposal for a toolkit approach to CDS and the remainder of the chapter elaborates on the key descriptive tools[1] that will be used to support the critiques offered in this book.

The critical in (multimodal) critical discourse analysis

Most accounts of the development of CDA attribute the influence of Marxism and Frankfurt School critical theory as central to its proponents' claim to being critical (Fairclough et al. 1997; Wodak and Meyer 2009; Wodak 2011). Authors such as Wodak and Meyer (2009: 7) directly link the project of CDA to the unfinished project of Enlightenment such that CDA is understood to be directly engaged in a broader emancipatory project of transforming social relations rather than merely studying them. As Fairclough (1995: 36) puts it:

> My use of the term "critical" (and the associated term "critique") is linked on the one hand to a commitment to a dialectical theory and method "which grasps things . . . essentially in their interconnection, in their concatenation, their motion, their coming into and passing out of existence" (Engels 1976: 27), and on the other hand, to the view that in human matters, interconnections and chains of cause-and-effect may be distorted out of vision. Hence "critique" is essentially making visible the interconnectedness of things.

Indeed, as methodologically and theoretically diverse as the CDA approach is, for Wodak and Meyer (2009: 6), it is the centrality of critique as praxis that binds CDA practitioners together. Being critical, then, is not simply an apolitical practice of applying rational thinking to matters at hand so as to circumnavigate errors and inefficiencies in thought and action, as it is now taught in many liberal arts programs. Instead, being critical is expressly a political act. Following from Marx's 11th thesis on Feuerbach, as a program of action, CDS seeks to not simply describe the representations of society but also intervene in those representations so as to transform society. Accordingly, CDS seeks to take on an advocacy role, not only challenging structures of social domination as they are encountered but actually championing those groups that are subjected to those structures of domination. Thus, much of CDS work has focused upon the representation of immigrants, the poor and working classes, women, and other marginal or subaltern groups. These representations are drawn from a wide variety of everyday and institutional settings so as to include news reports, magazine articles, advertisements, textbooks, music, government documents, etc. As a critical project, CDS seeks to reveal how macro-social relations of domination are enacted through, but also obscured in, the micro-politics of such texts.

By and large, most CDS researchers tend to see power and the structures that underwrite the reproduction of asymmetries in power relations as being obscured to those who are subject to them. As Fairclough describes it above, inequality is "distorted out of vision." Likewise, Fairclough et al. (2011: 358) propose that "CDA aims to make more visible these opaque aspects of discourse as social practice." In this way, CDS researchers have tended to understand the way people draw upon semiotic resources as being ideologically shaped in ways that are not transparent or self-evident. By engaging in CDS, it is thought that researchers will be able to bring these ideological workings to the surface, making them more obvious, and, presumably, making those who are subject to them more conscious of and able to critically respond themselves to those workings. At the same time, Wodak and Meyer (2009: 7) also seek to make it clear to their readers that in adopting the role of critic, researchers must be self-reflexive enough to recognize "that their own work is driven by social, economic and political motives like any other academic work and that they are not in any superior position." Nevertheless, the framing of the critical project of CDS as one of demystifying power relations that otherwise remain hidden to social actors raises some ethical-epistemological and, perhaps, even ontological problems for those doing the work of CDS.

In his essay, "Why Has Critique Run out of Steam?," Latour, very much redolent of Baudrillard's (1981) charge of the fetishism of fetishism, takes to task those who would take up the role of critic. Somewhat hyperbolic in tone, Latour (2004: 297) charges that "90 percent of the contemporary critical scene" can be summed up in what he terms the "fairy position" and the "fact position." The fairy position is occupied by those who "associate criticism with antifetishism." As the name would suggest, the antifetishist seeks to "show that what the naïve believers are doing with objects is simply a projection of their wishes onto a material entity" (Latour 2004: 237). For Latour, this puts the critic in an intrinsically privileged position in relation to those idolaters in need of enlightenment. At the same time, the critic is also able to enjoy the fact position whereby the naïve are now shown that unbeknownst to them, "their behavior is entirely determined by the action of powerful causalities coming from objective reality they don't see, but that you, yes you, the never sleeping critic, alone can see" (Latour 2004: 239). This unabashed, urbane self-privileging leads to what Latour (2004: 240) then calls "critical barbarity" whereby the critic simultaneously draws from three incompatible repertoires: "We explain the objects we don't approve of by treating them as fetishes; we account for behaviors we don't like by discipline whose makeup we don't examine; and we concentrate our passionate interest on only those things that are for us worthwhile matters of concern." In sum, for Latour, the critical barbarian is one who assumes to know better but is also, I will argue, something of a straw person.

In light of this, Latour (2004: 243) proposes a third position—the "fair position." While the unfair critic is said to claim a monopoly in sophistication and knowledge, to become a fair critic, for Latour (2004: 231), means "the cultivation of a *stubbornly realist attitude*" with a new set of descriptive tools that do more than simply "debunk" or reassert the privilege of the knowing critic. To Latour (2004: 246), the revitalization of critique is only made possible by taking up the fair position in which the critic does not negate, but he adds: "The critic is not the one who lifts the rugs from under the feet of naïve believers, but the one who offers participants arenas in which to gather." This strikes me as very much the kind of ethos that CDS researchers have sought to develop. Where CDS is entirely incommensurable with Latour's position is in the claim that "explanations resorting . . . to power, society, discourse had outlived their usefulness and deteriorated to the point of now feeding the most gullible sort of critique."[2]

As an admonishment of ethical failing, I find Latour's charge to be, as I have already suggested above, somewhat hyperbolic. Returning to Wodak and Meyer's point that CDA researchers must be self-reflexive and not imagine that they somehow stand outside the very same social and economic structures that they are examining, it strikes me that it would be terribly naïve, if not elitist, of CDS practitioners to presume that only they can approach texts critically. The politics of CDS is premised upon advocacy and not voluntarism and, on the part of CDS scholars, self-describing as "critical" signals the "intention to make their position, research interests and values explicit and their criteria as transparent as possible, without feeling the need to apologize for the critical stance in their work" (Wodak and Meyer 2009: 7). So while the critical barbarians, by contrast, must obfuscate their own position in order to claim mastery over "matters of fact," CDS researchers explicitly attend to what Latour (2004: 231) characterizes as "matters of concern." To attend to matters of concern, rather than merely matters of fact is, for Latour, to attend to those entangled interests and politics that are gathered into the objects/texts that we study. What Latour is calling for, therefore, is an empiricism guided by a politico-ethics of care and concern such that matters of fact emerge from matters of concern and cannot be used to ride roughshod over those concerns.

While I consider CDS as a program of research to be very much oriented to matters of concern, I believe that by framing our research as demystifying risks pushing us back onto the thin ice of matters of fact. Rather than understanding the critical practice of CDS to be one of revealing what is otherwise hidden to social actors, the term "de-naturalizing," as Machin and Mayr (2012: 5) use it, comes much closer to how we should frame the critical aspirations of a multimodal approach to CDS. My preference is for de-naturalizing because it suggests that discourse has done something or that something has been done with discourse that is not simply a concealment or a repression of reality. De-naturalizing suggests, instead, that something has been made to be experienced as natural and this is closer to Foucault's conception of the positivity of discourse and the productivity of power. Since CDS does take from Foucault the principle that discourses are constitutive, conceiving of discourse as a repression of the truth runs counter to that insight. Rather than characterizing the work of CDS to be a program of de-mystification, it would be more in keeping with its Foucauldian origins to, instead, understand the critical practice of "making visible the interconnectedness of things" (Wodak and Meyer 2009: 7) as that of describing how things (subjects, practices, facts, etc.) are made to be visible.

In Foucault's (1978: 12) well-known illustration of the positivity of power, he refers to the alignment of power with repression as "the repressive hypothesis." Using the stereotype of the prudish Victorians and their supposed banishment of all matters sexual as an illustration, Foucault argues that, in reality, there was an explosion of discourse on sex rather than a silencing. Rather than repressing sex and making it disappear, sex was, instead, organized into a positivity of power/knowledge that made matters of sex visible but in very particular ways: "There was a steady proliferation of discourses concerned with sex-specific discourses, different from one another both by their form and by their object: a discursive ferment that gathered momentum from the eighteenth century onward" (Foucault 1978: 18). The point is that sex was not made to disappear but, rather, it was made to appear in very specific, historically contingent ways. Accordingly, Rajchman (1988: 91) characterizes Foucault's work as histories of such visibilities whereby his work gives us "pictures not simply of what things looked like, but how things were *made* visible, how things were *given* to be seen, how things were '*shown*' to knowledge or to power—two ways in which things became *seeable*." With this in mind, the "making visible the interconnectedness of things" is a matter not of revealing that which has been repressed but rather to reveal how those interconnections are made, built up, and what they make visible as self-evident or natural.

This is not to say that critical discourse research cannot address and highlight omissions and gaps in the ways in which actors, actions, and knowledge are represented but, to remain true to Foucault's conception of discourse as being both socially shaped and constitutive, those meaningful absences should be understood as being made discontinuous to what has been made visible rather than as a hidden or repressed truth to be liberated. What Foucault (1990: 67) proposes as the principle of discontinuity calls attention to

> the fact that there are systems of rarefaction [which] does not mean that beneath them or beyond them there reigns a vast unlimited discourse, continuous and silent, which is quelled and repressed by them, and which we have the task of raising up by restoring the power of speech to it. We must not imagine that there is a great unsaid or a great unthought which runs throughout the world and intertwines with all its forms and all its events, and which we would have to articulate or to think at last.

This does not mean that omissions and silences cannot be addressed critically but rather, to the contrary, in addition to detailing what has been made self-evident, we should also attend to how in making some actors, actions, and knowledges seeable, a given discursive regime also disarticulates other actors, actions, and knowledges so as to render them invisible. Again, this goes back to the "productive effectiveness" (Foucault 1978: 86) of power and discourse whereby both visibility and invisibility are positive effects of power. Such a critical practice, therefore, "does not arise out of a desire on the part of researchers to liberate those who were marginalized in all of their particularity (or peculiarity) or to diagnose their reasons for so being, rather, it is to understand the wider historical discourses that operate to maintain their position at the edges or on the margins of society" (Anaïs 2013: 133). In this way, a CDS approach can be better said to make visible and de-naturalize how some things have come to be interconnected and other things disconnected. Accordingly, instead of a program of de-mystification, which depends upon a notion of revealing "a great unsaid," CDS research documents the productive effectiveness of discourses in creating the self-evident character and acceptability of a given set of practices. And, importantly, this process of documenting or describing discourse has critical-transformative potential because "to see the events through which things become self-evident is to be able to see in what ways they may be *intolerable* or *unacceptable*. It is to try to see how we might act on what cannot yet be seen in what we do" (Rajchman 1988: 94). In other words, in analyzing how discourses come to be realized in texts, making visible the interconnectedness of things, we have the opportunity to re-order those things into new connectivities and arrangements. As Fairlcough (2013: 7) puts it in his updated "General Introduction" to *Critical*

Discourse Analysis, "Critique assesses what exists, what might exist and what should exist on the basis of a coherent set of values." There is no easy clear-cut division between the negative and the positive in critique. The critical impetus in CDS, therefore, is not simply a matter of negation and dismissal. As a "problem-oriented, critical approach to research" CDS starts not with a "fixed theoretical and methodological position," (Wodak 2011: 40) what Latour would call starting from a position of "matters of fact," but "with a research topic that is a social problem," (Wodak 2011: 40) or in other words, a "matter of concern." In sum, CDA, multimodal or not, should be thought of as "discourse analysis 'with an attitude'" (van Dijk 2001: 96).

Description as a tool of critique

We frequently refer to work that is insufficiently critical as being too descriptive—that it merely details the phenomenon in question "as it is" without actually subjecting it to any sort of reflexive evaluation or analysis. Van Leeuwen (2012) has noted that in the emergent area of multimodality, there is a considerable body of work that tends, in my words, to be of a more celebratory bent and which "focuses more on heralding what is to come than on critiquing what is." Djonow and Zhao (2014) also note that while MDA and CDA share an understanding of communication as being both a social process and a multimodal phenomenon, MCDA is, by and large, still an emergent field of research. While some of the more "reverential" MDA literature is clearly caught up in what Evgeny Morozov (2014) calls "technological solutionism"—the tendency to jump to quick technological fixes without adequately researching and understanding what the "problem" actually is— there is, of course, a place for asserting what *should* exist. For example, in education and curriculum studies, multimodality has been taken up as the new solution to pedagogic issues surrounding the teaching of literacy (van Leeuwen 2012: 2). The problem is when researchers examine multimodality in a vacuum and reduce it to a set of formal properties and techniques rather than sets of resources contextually selected and used in meaning-making practices. So from the critical perspective, being descriptive, in and of itself, is not the issue—it is the failure to contextualize the knowledge gained through description with systemic analyses of social problems.

Machin also finds value in description. In attending to the elements that comprise a text, we must also consider the relationships between those elements. For Machin (2013: 2–3), there is "power" in description because a systemic approach to analyzing discourse in semiotic modes other than language "is useful for drawing out buried ideologies"; however, he also cautions that it is "possible for work to focus on being too technically descriptive" at the expense of critique. However, Machin's reservation is not only that in description, critique might be derailed, but also that the kinds of syntagmatic descriptive practices applied by CDS to language may not transfer so neatly to other semiotic modes. What Machin doubts is whether or not those relations between non-linguistic elements in multimodal texts can be treated as homologous with syntax in language. In other words, describing meaning-making practices such as photography or painting in terms of language can, in fact, impair our ability to analyze resources as they are *used*. What this means is that a critical approach to multimodal discourse analysis necessitates engaging in a systematic practice of describing how the text is composed, but in a manner that moves beyond thinking that all relations between elements within a text must be syntagmatic ones: "What, I would suggest, is important is not how the process is like or unlike language in this manner, but what the resources *do* and how can they be used as they sidestep and gloss over certain kinds of commitments" (Machin 2013: 5). To this end, Machin (2011), see also (Abousnnouga and Machin 2010: 138), proposes a toolkit approach starting with a two-step model of description/ analysis borrowed from Barthes' conception of denotation and connotation.

The denotation/connotation binary was first proposed by the utilitarian philosopher, John Stuart Mill who proposed that words both denote, or refer to, a class of things and connote or imply attributes associated with that word. So, for example, the word robot refers to a class of technical objects that operate using both a sensor and an actuator system for acting semi- or fully autonomously in their environment. At the same time, starting with the play, *R.U.R. (Rossum's Universal Robots)* by Karel Čapek (1920), in which the word robot is introduced, the word robot is also clearly used to connote proletarian labor. Barthes (1977) takes up this distinction through the work of linguist Louis Hjelmslev and applies it principally to photography. Barthes' (1977: 17) position is that as a representation, each text bears two messages: the denoted message and the connoted message. This, in effect, creates two levels or orders of meaning, much as he represents the relationship between language and myth in *Mythologies* (Barthes 1973). Barthes (1977: 17) characterizes the denoted message as "the analogon to itself" such that it serves as a direct referral to the actual thing while the connoted message calls upon all those abstract associations a society has attached to the referent.

Abousnnouga and Machin (2010: 138) rightly point out that, as a theory of semiosis, this is somewhat problematic since, despite what the concept of denotation seems to suggest, "we never really experience any part of communication in an innocent way." Accordingly, they propose that Barthes' denotation-connotation be adopted, instead, as a kind of guiding procedure for analyzing texts in which a rigorous approach to description supports and is not overshadowed by "our eagerness to show what something means" (Abousnnouga and Machin 2010: 136). Critical textual analysis requires attending to *how* a text is made meaningful as much as *what* the text means. By attending first to describing *how* semiotic resources are selected and articulated within the text and then using that description to detail *what* meanings those resources can afford, we are able to examine precisely the ways in which discourses once textualized can make visible and self-evident some versions of reality while obscuring others. Description then, when not understood as an end in its own right as in more formalist exercises, becomes a tool of critique when it is called upon to support the claims we make regarding what discourses are being accomplished in the texts that we analyze. So while denotation-connotation taken as a theory of communication in which meanings are produced through a two-step process does not serve us very well; taken, instead, as an analytical tool, a two-step method of description-critique affords a more systematic account of exactly how meanings are accomplished. It allows us, to borrow from Latour, to adopt a "stubbornly realist attitude" by providing a robust set of tools for empirically examining the constitutive elements of semiosis as they are realized in discourse. To this end, the social semiotic approach to MCDA, as typified by Kress, van Leeuwen, and Machin, offers us a good number of critical tools that can be drawn upon in order to methodically detail and systematically analyze multimodal texts.

The survey of resources below adopts the toolkit approach proposed by Machin (2010, 2011 and Machin and Mayr 2012) whereby semiotic resources are described as they are utilized as parts of option networks available to communicators and then analyzed according to the communicative functions that they realize. There is a certain pragmatism on my part in terms of what I have included in the toolkit and what I have put aside as being less germane to our purposes here. The intention is to provide an inventory of semiotic resources that can be readily drawn upon to explicate how texts accomplish the meanings we ascribe to them when concerning ourselves with matters of how the technological features in the constitution of identity, power, and agency. Though certainly not exhaustive, this is not the same thing as a haphazard approach since the resources I have chosen to address here all contribute in significant ways to the systemic production of meaning. Again, the aim is not to describe *everything* but rather to selectively describe those elements that make up the key representational, interactional, and compositional options available to communicators. It is a false exactitude to try to make the map equal to the territory since that makes description itself the objective of the exercise. Instead, if we want to know something of the territory, we need to be selective about which features should be made to stand out. In other words, to continue the cartographic metaphor, our maps should make visible the means by which particular visibilities are produced.

With that objective, then, the remainder of the chapter offers an inventory of resources that will serve as the toolkit for analyzing technological discourse in the ensuing chapters. Specifically, the inventory includes connotation and metaphor, transitivity analysis, interaction analysis, modality, color, typography, and layout. The first two, connotation and metaphor, have to do with what can be called lexical and iconographic choice. When examining the kinds of connotations and metaphors that communicators select, the focus is on what kinds of word and image elements have been made present and the motivations they imply. Transitivity analysis examines the system of resources that realize how actors and their actions are semiotized as representations. Interactional analysis, in turn, concerns how interpersonal meanings are invested through the manner in which those actors are represented. Modality also realizes interpersonal meanings but chiefly those between the communicators themselves. The last three—typography, color, and layout—are compositional resources and serve to textualize the representational and interactional meanings through the coordination and differential emphasis placed upon elements within the text.

Lexical and iconographic choice

The words and images we choose are meaningful in terms of the kinds of associations that they carry for us. Social semiotics understands communication as a process of resource selection in which communicators draw from available option networks to both produce and interpret texts. Not all options are necessarily available to individual communicators as the kinds of texts we can produce are constrained by the cultural, social, political-economic, and historical contexts in which they find themselves. Drawing from the ideas of Vološinov, we can claim that communication always takes place not between individuals but between participants within a socially organized "inter-individual territory": "It is a territory that cannot be called 'natural' in the direct sense of the word: signs do not arise between any two members of the species Homo sapiens. It is essential that the two individuals be organized socially, that they compose a group (a social unit); only then can the medium of signs take shape between them" (Vološinov 1973: 12). Signs, or rather, semiotic resources are collective and are selected and articulated in texts to produce meanings based upon their affordances. These affordances entail historically derived associations that are evoked through their inclusion in the text. The two primary ways in which such meanings are constituted through association with meaning are connotation and metaphor.

Connotation

Connotation, as I have already noted above, refers to the more abstract second-order meanings associated with elements within a text. While denotation suggests that an element within a text "is what it is," connotation calls attention to how elements carry or are attributed with meanings beyond the immediate or literal ones they denote. While the image of a CCTV camera can be just that, the image of an electronic device for the purpose of recording moving images of a public space, images of CCTV cameras are also commonly used to connote concepts such as surveillance and control, and the loss of privacy. As van Leeuwen (van Leeuwen 2005: 37) explains, there is the layer of denotation, that is the layer of "what, or who, is represented here?" and the layer of connotation, that is the layer of "what ideas and values are expressed through what is represented, and through the way in which it is represented?" Following Barthes, van Leeuwen (2005: 38) proposes that connotation is effected in two ways: through "associations which cling to the represented people, places and things" and through the "specific aspects of the *way in which* they are represented" such as photographic techniques—what Barthes (1977: 23) calls photogenia—like lighting, exposure, focal distance, and so forth. Of course, it needs to be reiterated that in characterizing the semiotic potential of elements within a text in terms of layers, it is not being proposed that semiosis actually follows a two-stage process. We do not engage with texts first on an "innocent," literal level and then on a symbolic level. Rather, connotation points to the potential for "things" or resources to take on and therefore afford meanings that are historically and culturally derived, but not necessarily intrinsic to them.

Connotation is, on the one hand, an example of how meaning is the production of convention much as Saussure treated it. Associations are culturally rather than individually determined since it is through collective use that connotations become decipherable and enduring. On the other hand though, connotations are also a means of semiotic innovation or change, as van Leeuwen (2005: 37) points out. Because Saussure tended to characterize semiosis as the product of a closed sign system, he tended to treat associated meanings as not only arbitrary but also inviolable. Social semiotics, with its emphasis on semiotic practice, instead treats connotation diachronically rather than synchronically. Accordingly, van Leeuwen, while recognizing that connotations are historically produced, also proposes that new connotations are created when they are "imported" into new domains so as to create new meanings that draw from but modify established ones. For example, offering the example of the airline uniform, van Leeuwen (2005: 40) points out how different elements of "uniformness" from established uniforms are imported into the flight cabin to create "a novel 'composite of connotations'." The resultant of mixing elements such as color, garment types, embellishments, and accoutrements denotes the uniform of airline cabin crew but also connotes "a complex of abstract concepts and values" specific to the corporate identity and branding of that airline. This practice of semiotic bricolage is very much akin to what Hobsbawm and Ranger (1983) have called "the invention of tradition" whereby new "traditions" are invented to legitimate the present through the appropriation of old ones in a manner that makes the past seemingly connect seamlessly to the present. What these two examples demonstrate is how social groups establish and mobilize their representations of reality by creating new associations through recombination and it is by drawing from already established convention that the new hybrid connotations are both generated and take hold.

Obviously, all semiotic resources are invested with connotative values. Communicators draw from, and articulate, them according to the abstract concepts, ideas, and values that they potentially realize. In visual communication, images that have more abstract qualities tend to be understood as being more expressly connotative. Machin's (2004) work on image banks proves an interesting case since the images are created to be open-ended in terms of use. Images are banked by concept and the more concept categories the image can be filed under, the more potentially profitable the image. As a result, stock photos are created to be "multi-purpose, generic and decorative" (Machin 2004: 317) and ready to be contextualized rather than documenting a specific image-event. This leads Machin (2004: 326) to speculate that "it seems, we are moving to a photography in which there is only connotation." While not all forms of communication emphasize the connotative function to the same degree, no semiotic resource can be purely denotative, since as Barthes (1977: 45) points out, it is the denotative aspect of a text that potentially obscures or "innocents the semantic artifice of connotation." Of course, stock photography is created to be inserted into and consumed within other texts. Once incorporated as an element within a multimodal text, the image is contextualized and its connotative potential is tied to the communicative event in which the image has been enrolled. The point is that if communicative events are produced through the selection of semiotic resources, those selections are made not simply to realize particular denotative tasks but are also motivated by the connotative potential of the individual resources.

One way to begin detailing the connotative potentials made available in a verbal text is to simply begin by looking at the kinds of words that have been utilized. By looking at lexical choice, we can begin to read between the lines and extrapolate how the subject matter of the text is being articulated in relation to discourses available to the author (Machin and Mayr 2012: 32–33). So, for example, the following is extracted from a newspaper article introducing readers to the Internet of Things concept.

"Internet of Things" [shares] data without your help; [Potential] for [smart] devices is ["boundless,"] says Waterloo chip developer
(LaSalle 2012)

[Smart] vehicles, home security systems and web-connected appliances are becoming part of what's known as the "Internet of Things," a growing network of objects [creating] and [exchanging] data *without people directly involved.*

Things are getting a ["voice,"] says Ric Asselstine, chief executive officer of Terepac Corp., a Waterloo company that makes tiny electronics to put into objects to make them ["smart"] and compatible with the Internet of Things. "At the end of the day, what we're [creating] is information," Asselstine said in a phone interview from Terepac's headquarters on Colby Drive.

There is the [potential] for "trillions" of devices to be connected to the Internet of Things, he said, noting all of the objects in his office alone. "The [potential] is literally [boundless]."

. . .

A [smart]phone can automatically [tell] you that you have email, update programs, [sense] your location, [talk] to other servers and back up information, said Tauschek, lead research analyst at the Info-Tech Research Group in London, Ont.

"It's [doing] stuff as it was programmed to do [without me] having to intervene," he said.

"They're devices and inanimate objects that [exchange] data [without, basically, human] intervention through a network, oftentimes over the internet."

However, there is a basic [human risk] with all of that <information>.

"If the [wrong person] gets access to that <data>, to that information, they can find out [potentially] <personal things> about your location. That's identity theft," he added.

. . .

The Internet of Things is a term believed to be first used in 1999 by British tech pioneer Kevin Ashton. He talked about computers being *able to use data they gathered without human help*, to track and count things to [know] when they would need replacing, repairing or recalling.

Asselstine takes the view that inanimate objects can [provide] useful information to people if they're connected.

"What if this could [talk?]" he asks.

The first thing to note is the frequent use of the word "smart" at the start of the news article. Since, by convention, the most important information is written into the first part of a news report and "lower value" contextual information offered at the end, the occurrence of the word in the title and the first two paragraphs suggests that this is to be crucial to understanding the significance of these devices. Smart in reference to technology, of course, connotes autonomy as well as artificial intelligence. We are told that the devices operate "without people directly involved." Furthermore, in the popular imagination, artificial intelligence is typically equated with symbolic artificial intelligence, which aspires to replicate human intelligence whereas there are, in fact, other more behavioral models of artificial intelligence that are probably more appropriate for describing "smart devices" (see Bradshaw 1997). This notion of an autonomous human-like intelligence is further reinforced by the kinds of words used to describe the processes that the objects are reported to perform: creating, exchanging, sharing, telling, talking, and sensing. The first three verbs suggest behaviors that are intrinsically social and the last three are closely associated with language and cognition. These sorts of lexical choices work to further anthropomorphize the devices being described in the article, making them seemingly function like little "people."

Additionally, the scope of the article is signaled by the choice of the word "potential" in the article subheading, which is a paraphrase of a quote taken from the interviewee. The potential, the future, and the uses of this technology are declared to be "boundless." The impression is that the technology having been introduced, it will open the door to limitless technological innovation and that smart devices will increasingly replace non-smart ones. If the potential is boundless, it also suggests there is no reason to stand in its way and that any further development will only be a beneficial refinement of what is, at the core, a positive innovation. The one proviso comes from the risk that is qualified as "human." The risk to users of the technology arises when "the wrong person gets access to the data." This risk then arises from "human nature" rather than something built into the technology itself despite the fact that it is patently a security and, therefore, also a system problem.

It is also worth considering how the word "potentially" is used here to modify the degree of risk users face. While in terms of the capacity of the technology to be adopted and further developed, its "potential" is "boundless," the risks associated with the use of the technology are diminished in significance by stating that should the "wrong person" access the data being circulated, "potentially personal" can be revealed. Here the choice of the word "potentially" modifies probability (see Halliday 1985: 335). While personal information could be wrongfully accessed leading to identity theft, it is rendered a possibility as if the data could be accessed and only non-personal information found instead. Since the reader's principal concern is likely to be the release of personal information to cyber-criminals (and not necessarily the actual collection of the information as data), framing the hijacked data as only "potentially personal" affords thinking that even in the event that security is breached, personal information may still be secure.

Finally, we should note that the lexical items "information" and "data" are used to describe what is circulated or "shared" between the devices. At the same time, though, what is not selected is equally meaningful. Omissions can, of course, entirely change how we understand what is being represented so, interestingly enough, there are no possessives used in reference to the data. We are told what kind of information becomes data but its ownership is left undefined, though obviously the information is coming from the consumer-users of the devices. Instead, we are told that it is the devices and the companies that are developing and selling them that are creating the data and information. So while the source of the data might be the behaviors of the device users, those behaviors are being transformed into a product called data, which is not actually the property of the users. Furthermore, the lack of possessives in relation to data and information also further diminishes the sense of risk one might experience since it is data that the wrong person might get access to and not *your* data.

Overall, the individual lexical choices made within the news item place the Internet of Things technology in a discourse on technology as progress. This is done by connotatively realizing specific statements about technology that are intrinsic to that discourse. First, technology is described as operating autonomously, independently of but also mirroring human actors. Technology, for good or bad, once created seems to take on a life of its own. Second, technology, once developed, has an inherent push toward improvement.

Metaphor

Metaphors are a special kind of meaningful association based upon a claim of semblance in which one thing is characterized as closely resembling another in some specific way. So, for example, computer security is often discussed in terminology associated with immunology such that computers are presented as being like organic systems vulnerable to infection and therefore capable of hosting foreign bodies referred to as viruses (1997). In *Metaphors We Live By,* Layoff and Johnson (Lakoff and Johnson 1980: 4) proposed that while most people assume metaphor to be merely a form of figurative or poetic language, metaphors are, in fact, "pervasive in everyday life" and rather than being simply "a matter of words," they are equally integral to thought and action. Metaphors are not simply flourishes or decorative devices but, instead, are crucial to how we come to define, experience, and act in our everyday lives. Our ability to understand and extrapolate depends, to a large part, on the ability to think about things in relation to other things that we already know. Accordingly, the use of the contagion metaphor shapes how we have come to understand and experience computers as relatively closed, homeostatic systems that are vulnerable to pathogens if not "immunized" rather than, perhaps, relatively open systems that operate in relation to other computational devices. In sum, metaphors are at the core of our conceptual systems and how we cognize the world around us. In this respect, metaphor becomes a site of intervention for CDA since "ideologically, . . . metaphor may be exploited in discourse to promote one particular image of reality over another" (Hart 2014: 137).

Van Leeuwen (2005: 30) contends that the "essence of metaphor is the idea of 'transference'" and traces the etymology of the word to the Greek word for transport. Ultimately, as van Leeuwen points out, metaphor is, in fact, a metaphor itself since it presupposes "a similarity between transporting goods between places and transporting words between meanings." Following from Lakoff and Johnson (1980), this process of transfer is from the source domain (the concept from which the metaphorical expression is to be transferred) to the target domain (the concept from which the metaphorical expression is to be applied), creating a new metaphor for conceptualizing and experiencing a given phenomenon. Metaphors, in effect, transfer a salient quality or attribute of a thing or action from one (source) domain over to a different thing or action in another (target) domain. So in the computer virus example above, there is a transfer from the source domain of immunology to that of the target domain of computer security. As a cognitive mechanism, metaphors enable our understanding of experience by drawing upon the established experience of another presumed similar domain. In this way, they are constitutive insomuch as they "create new meaning, create similarities, and thereby define a new reality" (Lakoff and Johnson 1980: 211). At the same time, however, metaphors also constrain the range of possible alternate ways of constituting experience by tying the experience being targeted to a specific already established reality. By drawing upon an immunology metaphor, the individual computer tends also to be conceptualized as an organism defined by discrete, bounded divisions that must be protected from transgression. Thus, "new metaphors will therefore highlight new aspects, but they may also obscure aspects which were previously out in the open" (van Leeuwen 2005: 32). As a primary resource for constituting reality, metaphor is essential to making things visible and self-evident by the creation of interconnections between source domains and target domains, "allowing new things to be noticed, new connections to be made, and new actions to be taken as a result" (van Leeuwen 2005: 32). Metaphors, then, afford particular shared frames of reference that can then be applied to the matter at hand, so as to enable a common experiential orientation between communicators.

The Mechanical Turk offers a good illustration of how metaphor can be used to capture and mobilize a particular way of experiencing technology. The Mechanical Turk was a device created in 1770 by Wolfgang von Kempelen at the height of the European automaton craze of the eighteenth century. While it was presented as an automated chess-playing machine with a mechanical turbaned player moving the chess pieces, in reality, it was an elaborate, approximately eighty-four-year-long deception. The "machine" could not operate independently and actually required a person to be hidden within the base cabinet of the device in order to operate. Seemingly a machine capable of mimicking human cognition by hiding its actual workings within the cabinet, the Automaton Chess Player has come to represent not just hidden labor but also the experience of being blind to that labor. Walter Benjamin (1968: 253) famously used the Turk as a metaphor for the relationship between more doxa-like variants of historical materialism and theology such that theology had come to be the hidden workings of historical materialism that his contemporaries had failed to acknowledge. More recently, though, the Mechanical Turk has come to be a metaphor for crowdsourced labor. After unsuccessfully trying to implement an Artificial Intelligence system for removing duplicates on the Amazon.com website, Amazon turned, instead, to human knowledge workers (Aytes 2013: 80). Calling it an Artificial Artificial Intelligence system, Amazon's Mechanical Turk was opened in 2005 as a public crowdsourcing site to contract freelance human knowledge workers to perform micro labor remunerated as Human Intelligence Tasks (Pontin 2007). Like its namesake, Amazon Mechanical Turk completes micro tasks in a seemingly automated fashion while obscuring the labor process that actually makes them possible. To "Turk it" has accordingly come to be used to refer to enlisting otherwise hidden "crowd" labor. Returning to the terminology of Lakoff and Johnson, crowdsourcing is the target domain that is conceptualized through the original Mechanical Turk as the source domain.

There is, of course, no shortage of metaphors that are applied to technology. The metaphors we use actively shape our experience and understandings of technology in general and how it is embedded in our everyday lives. One such common metaphor is "technology is a double-edged sword."

> For many couples, **technology is a double-edged sword**. The "his" and "hers" towels have been replaced by smartphones that allow people to stay tethered all day, whether it's to share shopping lists or heart-shaped emoji. But those same couples get into tiffs when one person pulls out a cellphone at dinner or clicks on the iPad before bed, forgoing pillow talk for Twitter. (Bilton 2014)
>
> What we do with these technologies is not preordained . . . I am optimistic that we will get more promise than peril. But they both exist. **Technology has been a double-edged sword** ever since fire and stone tools. (Hume 2014)
>
> Mobile connectivity can be **a double-edged sword: Technology** gives us the freedom to abandon the office and work anywhere; it can also be an electronic leash binding us to work anywhere, anytime. (Shaw 2013)
>
> The increasingly permeable boundary, supported by **technology**, between work and home **is a double-edged sword**. (Skinner 2013)
>
> They're supposed to have those PowerPoints ready by morning, but they're using those same laptops to write their novels. Point is, the way we work is changing, yes, and we're still unsure what the outcome will be. But it's that **double-edged sword of technology** that's behind it. (Craig Kirchoff 2012)

By attributing to technology (target domain) the quality of a double-edged sword (source domain), authors are able to talk about technology as having both positive and negative consequences—the idea that it cuts both ways. At the same time, however, this metaphor also affords thinking about technology as somehow intrinsically neutral. As Hume (2014) puts it, the consequences of a technology are not "preordained." What really matters in the end is how you use the technology or which way you swing the sword. Swords do not cut people, people do. As already discussed in Chapter 1, this is a deeply problematic assumption about technology that will be discussed in the next chapter. Nevertheless, while not breaking with the assumption of neutrality, the metaphor is extended by Craig Kirchoff (2012) to suggest that the outcome of adopting a technology cannot be fully anticipated and that it could, in effect, go either way. Also interesting is the use of the metaphor in the open letter against Killer Robots by Clearpath Robotics Cofounder & CTO, Ryan Gariepy.

> The Double-Edged Sword
>
> There are, of course, pros and cons to the ethics of autonomous lethal weapons and our team has debated many of them at length. In the end,

however, we, as a whole, feel the negative implications of these systems far outweigh any benefits. (Gariepy 2014)

When technology is qualified as a double-edged sword, it suggests that technologies have the potential to do either good or harm but are themselves neutral.

Metaphors can, of course, be realized in semiotic modes other than language. What matters is that a material semiotic resource is able to afford "a meaning potential that derives from our physical experience of it, from what it is we do when we articulate it, and from our ability to extend our practical, physical experience metaphorically to turn action into knowledge" (van Leeuwen 2006: 146). The poster image (see Figure 3.1) for an e-waste collection event employs such an experiential metaphor in order to establish a relationship between e-waste (target domain) and defecation (source domain).

Figure 3.1 *Poster for an e-waste collection event.*

The image is of a humanoid robot squatting and over a pile of e-waste. The squatting position is one that is most suited for human beings to release the contents of their bowels. And the pile of e-waste is, indeed, reminiscent of a pile of excrement. We can imagine that, having released the e-waste from the orifice at the bottom of its torso, it has undergone some sort of experience of relief, based upon our own experience of relieving ourselves of bodily waste. At the same time, as this is a multimodal text, the lexical choices and the typography also need to be considered in order to fully comprehend the meaning potential of the poster. The first line "Potty train your e-waste" functions as a command or injunction but also, given that the e-waste has already been expelled, an admonition. The remaining verbal text can be understood as an "offering" of information regarding the name of the event, its date, location, and website along with an additional command to "bring your old electrical and electronic appliances" to the location. The lack of curves in the typography selected for "POTTY TRAIN YOUR E-WASTE" and "ELECTRO RECYCLING_" metaphorically signifies reading from an analog video display screen and carries connotations of electronic artifacts. At the same time, while both are capitalized, the meaning potential of the capitalization differs. "ELECTRO RECYCLING_" is likely to be construed as being in title case since it is the name of the event, but the capitalization of the "POTTY TRAIN YOUR E-WASTE," as an injunction and in an electronic-style font, has a very different meaning potential. In electronic communications such as email, the longstanding convention has been to attribute shouting to text that appears in all-capitals. There is a metaphorical relationship between the use of capitals in electronic communication and volume in spoken communication. This is likely in part because of the few affordances of early electronic communication where users could not readily change fonts or font size, but there was also a kind of instrumental aesthetic that discouraged the use of flourishes at the expense of speed. For some communicators, even the use of a capital at the start of a sentence seemed to be construed as the adding of an unnecessary keystroke. By this convention then, "POTTY TRAIN YOUR E-WASTE" is to be experienced as a forceful, authoritative command. Admittedly, the actual wording seems a bit odd—you train the body, not the waste—but, nevertheless, the text demands that the viewers of the poster take responsibility for their e-waste. Furthermore, multimodally, the poster also rebukes/shames the viewer for presumed past failings by visually contextualizing the text with the defecating robot. As such, there are, in effect, two transferences taking place here. Clearly, potty training, learning to do the controlled and socially appropriate release of bodily waste, is being metaphorically connected to the proper disposal of electronic appliances. But also, by implication, the experience of shame/embarrassment at having soiled oneself inappropriately is being transferred from traditional potty training to e-waste training, thus making this a rather moralizing as well as instructive message.

Transitivity analysis

Transitivity, as discussed already above, is the system that realizes the representation of social actors and their actions. Transitivity structures thus comprise three potential components: the process or action, the participants, and the circumstances under which the process takes place. Detailing the realization of transitivity structures allows the researcher to highlight not only the way the actions of actors come to be represented but also the distributions of agency among participants. Thus, as a descriptive tool, van Leeuwen's (1995, 1996, 2008; see also Machin and van Leeuwen 2005) analysis of the representation of social actors and action brings our attention to the ways in which agency is assigned to participants of social practices. Just as there is no communication free of connotation, there is no neutral way to represent people or their activities. So the representation of who does what to whom and under what circumstances is of great interest to critical discourse analysts since asymmetries in power relations are deeply implicated in how transitivity comes to be semiotized. Of course, there is no necessary correspondence between the roles social actors perform in social practices and the grammatical roles that they occupy within representations of the activity in question. Instead, in semiosis, it is quite possible to "reallocate roles, rearrange the social relations between participants" (van Leeuwen 2008: 32) and, therefore, redistribute agency in accordance with the interests and purposes of the author(s). Van Leeuwen's approach to the representation of agency is useful since, though it remains very much based upon linguistic analysis, it is the sociological categories and the corresponding sociological actor that he privileges rather than those formal categories of linguistics itself. At the same time, the extensive inventory of options for semiotizing social actors that Van Leeuwen provides goes beyond our needs here, so, in keeping with our toolkit approach, we can get to the matter of things by taking our lead from the less embellished treatment that Machin (2011) derives from van Leeuwen.

Social actors

As van Leeuwen (2008) notes, in any given representation of social action, not all participants or social actors may be included. While exclusions may happen for benign reasons such as when there is an assumption that the exclusion is obvious and "goes without saying," it is equally important to remember that the inclusion or exclusion of social actors may well be motivated so as "to suit their interests and purposes in relation to the readers for whom they are intended" (Van Leeuwen 2008: 28). Such exclusions can take one of two primary forms according to van Leeuwen: suppression and backgrounding.

Instances of the first form of exclusion, suppression, occur when the social actor has been excluded entirely and there is no actual reference to the actor to be found in the entire text (Van Leeuwen 2008: 29). So, for example, the following statement suppresses the actor(s) responsible for actually dropping the cluster munitions:

> Some 1,228 cluster bombs containing 248,000 bomblets were dropped during the initial stages of Enduring Freedom.

Had the statement instead been presented as:

> Coalition forces dropped some 1,228 cluster bombs containing 248,000 bomblets during the initial stages of Enduring Freedom,

the actors responsible for deploying cluster munitions would have been clearly identified. In this way, what is not represented can also be meaningful and the absence of particular actors can entirely change how we understand what is being represented. The first statement leaves it open as to who is responsible for the use of cluster munitions. One can easily imagine that the munitions were used by both sides or perhaps only by the Iraqi forces since, in the lead-up to the war in Iraq, there was repeated reporting in the media that Saddam Hussein's forces had previously used chemical weapons indiscriminately against Iraqi citizens in Halabja. In the second statement, by contrast, the actor(s) responsible for using the munitions is clearly identified.

The other form of exclusion, backgrounding, occurs when there is some reference to the social actors elsewhere in the text but their role in the represented action has been de-emphasized (Van Leeuwen 2008: 39). Backgrounding can have the effect of diminishing the connection between a reported event and the actors that take part in the event without completely obfuscating culpability. The use of backgrounding can be seen in the following text (Keen 1991) taken from an early news report covering the US bombing of the Amiriyah shelter that killed more than 400 civilians in the first Gulf War.

> U.S. troops here are troubled by civilian deaths in the bombing of a Baghdad bunker, but they hope people back home blame Saddam Hussein, not them.
>
> "You've got to read between the lines," says Army Staff Sgt. James Gouge, 56, of Elizabethton, Tenn. "If there were civilians in that bunker, it's because Saddam wanted them there."
>
> U.S. military commanders say the bunker, once used as a civilian bomb shelter, had been functioning as a military communications center for three weeks.

ANALYZING MULTIMODAL DISCOURSE: A TOOLKIT APPROACH

Iraq claims hundreds of civilians died when two U.S. bombs pierced the building's roof Wednesday.

Many troops here have seen TV footage of civilian bodies being removed from the site, and they know it's hard for people back home to accept the grisly scenes—and the thought their husbands, fathers and friends are responsible.

In the statements above, we are told that US troops are troubled by the civilian deaths and that the deaths resulted from a bombing, and even that members of the American public might blame them. However, we are not explicitly told that US forces deliberately bombed the target. This is a good example of backgrounding since it is not until the end of the excerpt that we are told that US military personnel could be thought responsible for the tragedy.

In representing social actors, it is possible to represent them in different ways in degrees of specificity and generality. Actors can be represented as specific individuals (individualized) or as part of a group (collectivized). In the case of individualization, social actors are verbally and visually represented as distinct entities. In verbal texts, means of individualization include nomination, referring to the actor by name, and the use of honorifics, attributing special functional status to an actor such as "CEO," "Doctor," or "Captain." In visual texts, individualization is not so cut and dry, but generally the more an actor is made visually salient and distinct, the greater the degree of individualization, though it is entirely possible, depending upon context, for the individual to be treated as being representative of a "type."

The following text is taken from the caption accompanying a photograph posted on the UK Ministry of Defence Flickr.com photostream (Fox 2013). It is a photo of a group of Afghan men seated in a circle taking part in a shura. One man is standing addressing the group and is presumably the district governor.

A District Governor in Afghanistan conducting an outreach shura in Saidabad to coincide with the opening of a bazaar.

British and Afghan troops have continued to heap pressure on insurgents throughout the winter in over eight-weeks of daring operations in Central Helmand.

The joint operations were launched to clear a stubborn insurgent hotspot but recently ended in success following a massive engineering effort mounted by British and Afghan troops—at times while under enemy fire.

In this case, the Governor is individualized in the photograph by virtue of being seen to be standing up and speaking to the group of men. The seated men are visually collectivized as part of the shura participants. The caption does not refer to the actual men but they are implicated in the reference to the consultation meeting taking place. They are, in effect, collectivized by being backgrounded as "an outreach shura." At the same time, by using the indefinite article "a" in front of the honorific title, the individuality of the District Governor is diminished so that he is verbally collectivized as part of a class of local officials. Visually, then, the Governor is individualized, but the verbal text works to categorize him.

Representing social actors so that they come to represent a particular class or type is termed categorization. Whether individualized or collectivized, Machin (2011: 119) explains, represented actors can also be categorized. Categorization differs from collectivization in that the actors are not merely grouped and homogenized but connotatively function as representatives of a specific category of actor. Social actors can, therefore, be represented as both individualized and at the same time categorized by representing an individual as standing in for a particular type. In this case, the District Governor leading a shura, dressed in traditional attire, is accordingly categorized as part of the local government. Machin further elaborates that, in visual texts, categorization can be "cultural" or "biological" or a blending of both. In other words, visual categorization occurs along a cultural-biological cline with cultural artifacts such as clothing, hairstyle, and personal effects affording cultural categorizations and stereotyped physical features affording biological categorizations of actors.

In Figure 3.2, a US officer sits among a group of Afghan civilian men in a shura. Identified by name, in the verbal text and facing the camera in the photograph, the soldier is individualized. Excepting the uniformed man to the left of the US officer, all of the other men are attired in civilian clothing. The man on the left wears a uniform that differs from that of the named officer, and his face and most of his body are obscured, so his significance in the photo has been minimized. The other men, dressed as they are, can be categorized as Afghan civilians. The two men directly facing the officer wear turbans and the one on the left also wears a patu. Drawing from other media images of Afghan people, to western viewers, these clothing items have come to readily connote the cultures of rural Afghanistan.

In addition to examining how actors are represented, it is equally of interest to consider the kinds of roles that they have been allocated in the represented social action. Actors can be represented to varying degrees as active or passive. Accordingly, an actor is said to be activated when represented as an active agent, and passivated when represented as undergoing or at the receiving end of the process or activity being depicted. In the example immediately below, "Consumers" clearly functions as an activated participant

Figure 3.2 *DoD photo by Staff Sgt. Brian Ferguson, US Air Force/Released. https://www.flickr.com/photos/39955793N07/5277576119 US Air Force Capt. Ryan Weld, the Zabul Provincial Reconstruction Team intelligence officer, talks with Afghans from Khleqdad Khan village during a shura, or meeting, December 19, 2010.*

while in the subsequent sentence, "consumers" is activated but then as the sentence continues "them" and "they" are passivated.

> Consumers want targeted, value-added services and greater control over their energy use, but don't always trust their existing provider to deliver.

> It gives consumers more options and makes them feel they are getting the best value.

Moreover, passivated actors can be represented as either beneficialized or subjected. When an actor is represented as being at the receiving end, either positively or negatively, the actor is described as beneficialized. So, for example, in the following sentence, consumers are represented as at the receiving end and will benefit with "simple, secure access."

> We believe we have the technology, expertise and reach to bring simple, secure access to consumers worldwide.

And when an actor is given the status of object in the representation, the actor is described as subjected as in the first part of the following sentence.

> A Customer Engagement Solution enhances that process by bringing consumers into the loop and presenting them with their smart grid data in sensible, easy-to-interpret charts and tables, while giving them the tools to understand and actively participate in their consumption demands.

Initially, "consumers" are represented as being subjected or object-like since they are to be brought "into the loop." In the remainder of the sentence, they function as beneficiaries, receiving data and tools, while the text, ironically, explains how the consumer is being made to be an active participant.

Social action

Following from considerations of the agency allocated to actors, the kinds of social action associated with the represented actors are equally of concern. Like the representation of participants, social action is also subject to representations that can vary in the degree of congruence between the semiotic resource options of the text and the actual social practices as they are experienced. This is not to say that there is an unmediated and therefore recoverable reality outside of the text but rather that, as part of the practice of CDA, we can explore how the text constitutes a given social practice and which social actors and interests are best served by that particular representation. Accordingly, as a given social action is (re)semiotized, it can be represented either as being the product of human agency or as occurring independently of or autonomously to human actors (van Leeuwen 2008: 66). Van Leeuwen terms the former form of representation as agentalized (in other words, caused by human agency) and to the latter as de-agentalized (caused by forces outside of human will). This is largely accomplished through the kinds of processes that are used to represent social action.

The way we represent social action can, Halliday has proposed, be divided into six primary types of process: material (processes of doing), mental (processes of sensing), relational (processes of being), behavioral (processes of physiological and psychological behavior), verbal (processes of saying), and existential (processes of existing or happening). Material processes represent processes of "doing" something in the world. They usually entail performing actions that have a material consequence such as "Susan kicked the ball" or "He bought some popcorn." Mental processes represent processes of "sensing" such as thinking, feeling, or perceiving. In transitivity structures that feature mental processes, there is always one conscious participant who senses. So, for example, "She saw what happened" and "He liked the picture" are typical of mental processes. Relational processes represent states of being such as "The child is sweet" or "Andrea is the winner" and entail representing the subject in question in terms of an attribute or identity. Behavioral processes represent those actions that do not have material consequences in the world. These are typically psychological or physical behaviors such as watching, tasting, breathing, smiling, coughing, and laughing. Grammatically, behavioral processes fall somewhere between material and mental processes. The behaver is a conscious being but the process functions more like one of "doing" than "sensing." Often, this kind of process has one participant only and cannot be "transactional" in the way that material processes often are. To breathe on someone, for example, would be a material process (to force one's exhalation on another) rather than a behavioral one. Verbal processes represent symbolic exchanges of meaning in which an actor functioning as "sayer" relays some sort of "verbiage" or communication, optionally, to a participant functioning as a "receiver." The sayer can be non-human as in the case of "The sign says the lecture is cancelled today." Finally, existential processes simply represent that something exists or happens. In this type of process, the participant is an "existent," which/who is presented or depicted as simply being in a particular state of existing: "There once was an amiable guinea pig." The "existent" can be a phenomenon of any kind and is often an event, but nominalized, in effect becoming a "thing." So, for example, "There was an attack by rebel fighters last night" semiotizes attacking as an event and those doing the attacking into an optional circumstance of the event. Alternatively, in a material process an event is still a verb: "The rebel fighters attacked last night." The difference then is that in an existential process, the event becomes a "thing" whereas in a material process, the event is represented as a performed action. As a schema for describing the types of processes being realized in texts, as in the verbal examples above, we can then use it to document how the actions and activities of actors are realized in a way that affords greater or lesser degrees of agency.

Interaction analysis

Since texts function as exchanges as well as representations, the manner in which interactants are depicted in texts also semiotizes social relations. Interactional analysis entails first describing which semiotic resources have been selected and how they are articulated so as to produce interpersonal meanings. In verbal texts, interactional meanings are realized linguistically through the ways in which the listener/reader is positioned in the exchange and oriented toward the represented participants. For example, look to how "children" are represented in the following sentence:

> By being connected ourselves, we can ensure our children are safe in cyberspace, and benefiting from the new technologies.

As an offering of information, the reader is being addressed through the collective plural pronouns "our" and "we," which suggests that the relationship between reader and author is being presented as one of close affinity. Note also that the participant, "children," is preceded by "our," making them the collective possession of the author and the reader. As an exchange then, the reader is being addressed as being on the same side as the author and likely part of a much broader "we-community" (Schutz 1967), concerned for the safety of "their" children. So not only can we discuss what the plural possessive pronoun "our" does in terms of the agency of the children, but we can also consider how it semiotizes the relationship between presumed parent (reader) and child as one of caregiver, but also of authority.

Visually, interaction analysis focuses upon viewer-participant orientation. When analyzing images, as in the vacuum ad in the previous chapter, we examine three primary semiotic resources for representing actors that are implicated in the constitution of social relations between the interactive participants, the represented participants, and the represented participants and the interactive participants: contact, distance, and point of view. Contact refers to the kind of interaction or exchange between represented and interactive participants that is realized in the text. Depending upon whether or not the represented participant(s) returns the look of the viewer, the image is said to either demand (gaze returned) or offer (gaze absent). A participant is depicted with a "demand" gaze is suggestive of being in a position to interact and require something on the part of the viewer. Conversely, a participant depicted with an "offer" gaze suggests that the participant is being made available to but cannot require anything of the viewer. Distance refers to the social distance between the participant(s) and the viewer and is largely realized through the size of the framing of the participant(s). Close-up images connote familiarity and intimacy whereas long shots connote distant and impersonal social relations. Finally, point of view realizes relations of involvement and power and is largely accomplished through the angles of vertical and horizontal perspective. It is important to remember, however, that these features are part of a simultaneous system that functions together to produce the interpersonal meanings rather than additively (Kress and van Leeuwen 2006: 148).

The way in which these resources function in combination can be illustrated by applying interactive analysis to this image taken from a website promoting digital whiteboards:

Figure 3.6, is a photograph of students and a teacher engaging in a learning activity using an interactive whiteboard device. From Kress and van Leeuwen's schema, this would be considered an "offer" image-act since none of the represented participants directly face the viewer. This would connote that there is little "symbolic 'contact', 'interaction' between the viewer and the people depicted" (Machin 2011: 117) and that they are present more for our examination than interaction. In other words, they are being offered to the viewer rather than demanding something of the viewer. The camera-subject

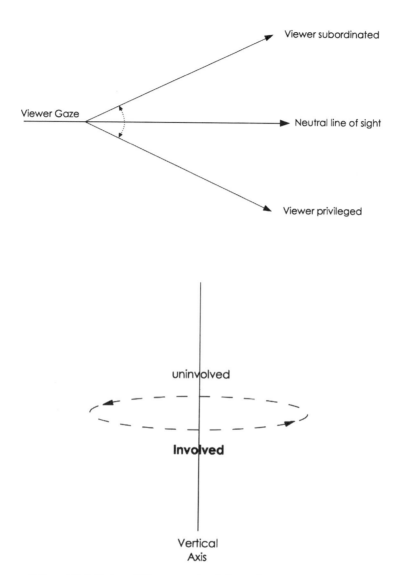

Figures 3.3 and 3.4 *Point of View as realized through vertical and horizontal axes.*

distance is close to medium with the students in the foreground and the teacher further back. This tends to structure the viewer's relationship with the represented participants so that the viewer is most involved with the students and then the teacher. At the same time, if we consider the horizontal angle of interaction, all of the represented participants look to the white board though the teacher is positioned at more of an oblique angle so that while she looks to the board, she is almost facing the viewer. She is, in effect, between the board and the students. This suggests a high degree of solidarity between

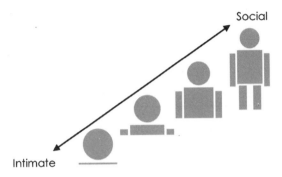

Figure 3.5 *Distance as a continuum from intimate to impersonal.*

Figure 3.6 *Digital classroom. http://education.smarttech.com/en/products/smart-board-800 courtesy of SMART Technologies.*

the represented participants, but it also separates the teacher somewhat from the students. Indeed, the accord between participants in the learning exercise is further affirmed through mirroring of the raised hand gesture. The teacher also does this, but she does not raise her hand as high as the two students. The choice of vertical angle places the viewer on the same plane with the teacher, thus suggesting an equality, but it also places the students at a slightly lower angle and so privileges the viewer. The angle is not pronounced and it is worth noting that the teacher is pretty much at level with the students. These vertical angles are thus very much in keeping with current discourses on pedagogy in which there is to be a kind of "democratic" relationship between students and their instructors based upon facilitation and expertise rather than bureaucratic authority. Together, the three resources semiotize social relations between viewer, students, and teacher as well as teacher and students in a way that conveys those kinds of more lax relations associated with collaboration and facilitation in active learning.

At the same time, on the webpage, the image is also accompanied by this superimposed text:

Intuitive at every touch

Go from pen to fingers to palm of your hand, interchangeably with object awareness. Flick, rotate, zoom and erase with natural touch gestures.

Interestingly, the text focuses upon the interaction between user and device rather than the kinds of personal interactions depicted in the image. As a multimodal text, the image functions to highlight the kinds of classroom interactions the learning technology will foster while the text is directed more at the individual user, promising ease of use. This is very much in keeping with traditional conceptions of the division of semiotic labor between language and images. Images are drawn upon to represent the emotive while language is drawn upon for the technical-rational. By using both semiotic modes, the manufacturers offer this image in order to promote their classroom technologies as being student-centered, easy to use, and "a catalyst for classroom participation." Therefore, it is not surprising that the image highlights the human interactions and backgrounds the technology itself, leaving it to the floating text, to offer more practical-technical kinds of information about actual use.

Modality

Modality refers to the social semiotic approach to the representation of truth. It differs from philosophical and logical matters of truth in that our concern is not "how true is this?" but "how truly is this represented?" Modality then reflects the text producers' level of commitment to what they represent as real (Machin and Mayr 2012: 186). As Halliday (1985: 335) puts it, modality "refers to the area of meaning that lies between yes and no." Semiotically, at least, truth is a matter of degree.

Modality is considered to be an interactional rather than a representational resource because "modality cues" function to produce "shared truths aligning readers or listeners with some statements and distancing them from others. It serves to create an imaginary 'we'" (Kress and van Leeuwen 2006: 155). In other words, modality establishes an orientation toward what is being represented based upon the proposed relation between communicators rather than the fidelity of the representation itself. Therefore, modality is, first, the degree of truth as the interactants understand it and, second, the resources they opt to use in order to express that claim to truth.

Modality is realized linguistically in terms of certainty. As Halliday (1985: 332) explains, modality realizes the writer's "opinion regarding the probability that his [sic] observation is valid." So modality can be expressed using high medians and low modals:

Science is the path to truth. : high modality

Science is probably the path to truth. : median modality

Science might be the path to truth. : low modality

The first sentence expresses a high degree of certainty on the part of the speaker while in the following two sentences the speaker's certainty is progressively diminished by the choice of modals. So modality is not a measurement of how true a statement like "Science is the path to truth" actually is, but rather the degree to which the truthfulness of the statement is actually asserted. In answer to the question "How truly is this represented?" the first sentence represents the claim that science is highly truthful and, therefore, with a high degree of modality while the second and third represent the claim with increasingly less certainty and, therefore, increasingly lower modality. What this example demonstrates is how, in everyday communication, claims to certainty can be modulated.

Likewise, as exchanges between communicators, images also carry modal cues. Just as in language, when examining a visual representation, it can also be asked "How truly is this represented?" In visual forms of communication, modality is realized through the modulation of detail, background, depth, color, and tone. Detail is realized along a continuum from maximum to minimum, with decreasing detail suggesting greater abstraction. Similarly, background can also be scaled on a continuum of varying degrees of detail. The less detailed the background appears, the greater the sense of decontextualization. Depth ranges between deep and flat or 2-dimensional perspectives. Naturalistic images are experienced as high modality when they fall on the deep perspective end of the continuum, but technical drawings often employ an isometric perspective and are also experienced as realistic renderings with only a median articulation of depth. Color modality markers are realized through the articulation of saturation (full color to black and white), differentiation (polychromatic to monochromatic) and modulation (multiple hues to flat). Finally, articulation in degrees of tone varies between high (sharp) and low (soft) contrast. Again, in photo-realistic images, tone tends to be high contrast whereas in more impressionistic images, contact tends to be lower.

There is, of course, no singular form of modality that is understood as being truthful for all parties in all contexts. Drawing from the work of Bernstein (1981) on coding orientations, Kress and van Leeuwen (2006: 165–66) offer, instead, four distinct types of visual modality: technological, sensory, abstract, and

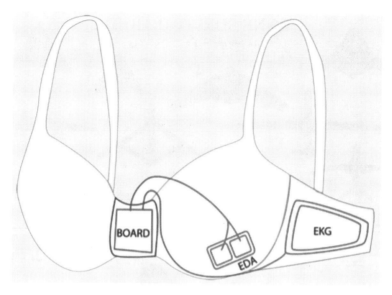

Figure 3.7 *High technological modality. Image courtesy of Mary Czerinski, PhD.*

Figure 3.8 *High sensory modality. Image courtesy of US Department of Agriculture. https://www.flickr.com/photos/usdagov/16379727731/.*

naturalistic. Technological modality is characteristic of schematic diagrams and blueprints and its veracity lies in its effectiveness and practicality. Figure 3.7, for example, depicts the prototype for a "smart-bra" that is designed to monitor heart rate and alert the wearer when emotional overeating is likely to occur. Sensory modality is characteristic of images intended to invoke corporeal pleasure or displeasure such as you would find in a cookbook or in popular nature photography (see, for example, Figure 3.8). Abstract modality is realized in visuals that are intended to convey the common or essential qualities of the subject matter rather than details particular to the specific subject rendered in the image. These are typical of abstract forms of art and also visuals such as you would see in a museum where the basic form remains naturalistic but details deemed unnecessary for comprehension are omitted. Here the

Figure 3.9 *High abstract modality. http://www.dronesurvivalguide.org/.*

emphasis is upon revealing the deeper essence or truth of the subject rather than, say, responding to pragmatic need as in technical modality. Figure 3.9 is a collection of drone silhouettes from the Drone Survival Guide, which, as a critical-creative project, allows (albeit tongue-in-cheek) the layperson to identify drones by their basic form. Finally, naturalistic modality corresponds with photo-realistic representations of reality such as in nineteenth-century landscape paintings or much of photojournalism. In this coding orientation, the more an image resembles how we would see something in real-life conditions, the higher its naturalistic modality. The image of the Bradley Fighting Vehicles in Figure 3.10 is a good example of high naturalistic modality with its high depth, detail, and natural hues. These, it must be remembered are types, and, in fact, modalities themselves can and often are articulated and mixed in images (see Machin 2011: 59–61). Figure 3.11 represents a good example of such a mixed modality image. The foreground is naturalistic but the glare from the sun is abstract, giving the image an almost artistic quality that has become typical of much military photography.

Typography

Van Leeuwen (2004, 2005, 2006) has observed that historically, typography has been regarded as being concerned with matters of presentation, but actually

Figure 3.10 *High naturalistic modality. US Army photo by Spc. Bryan Willis.* https://www.flickr.com/photos/soldiersmediacenter/6237767019/.

Figure 3.11 *Mixed modality. US Army photo by Sgt. Michael J. MacLeod.* https://www.flickr.com/photos/soldiersmediacenter/7257656984/.

not meaningful in its own right. This, of course, is another variation of the conduit metaphor whereby meanings are to be channeled with a minimum of distortion or interference via a "neutral" medium. For example, van Leeuwen cites McLean's *Manual of Typography* (in van Leeuwen 2005: 28), which acknowledges that the choice of typeface can "to a very limited extent . . . help to express a feeling or a mood that is in harmony with the meaning of the words" and thus is restricted to playing a supporting role to language rather than being meaningfully constitutive in its own right. The role assigned to typography then was to transmit the words of authors as clearly and legibly as possible without "deforming" or adding to the meaning of the text. However, van Leeuwen (2006: 139) posits that there has been a shift in the thinking and practice of typography such that it is increasingly implicated in "the cohesive work that used to be done by language." Furthermore, this trend extends beyond professional design practices to writing in general (van Leeuwen 2006: 142). To this extent, typography is clearly being utilized as a semiotic resource to contribute to the meaningfulness of the text as a whole:

> Typography is vitally involved in forging the new relationships between images, graphics and letterforms that are required in the age of computer-mediated communication, a form of communication which is on the one hand far more oriented towards writing than previous screen media such as film and television, but on the other hand also far more visually oriented than previous page media such as books. (van Leeuwen 2005: 29)

As a semiotic resource, then, typographic choice realizes meanings in relation with the author's lexico-grammatical choices. Furthermore, according to van Leeuwen (2006), typography is able to realize all three forms of meaning and therefore functions as a semiotic mode.

As a semiotic mode, typography functions as a system capable of realizing all three metafunctions. Representationally, typeface contributes to the representation of what is taking place by representing qualities or actions (Machin 2011: 89). For example, in Figure 3.1, consider the choice of a typeface that resembles the graphical display of an old low-resolution CRT-type monitor used in the e-waste poster. On the one hand, the blocky, pixellated style of typeface is often used to connote qualities of the electronic since it differs markedly from the curved, near letter-quality typefaces contemporary computer display systems are capable of rendering. At the same time, the difference from screen fonts today as well as the one used to display the details of the event also affords a sense of obsolescence. This is underscored by the use of the caret-style cursor used in text-mode computer displays, which appears at the end of "RECYCLING." Typography also realizes interactional meanings by expressing moods and attitudes toward what is being represented by the text. Returning to the Electro Recycling poster, we see two lines presented in the jaggie-style computer typeface. The first one presents the demand speech act, "Potty train your e-waste" and the second presents the offering speech act, "Electro Recycling." In the first speech act, there is a demand for service while in the second, there is an offering of information. The offering of information is presented in a larger scale than the demand. This would suggest that the issuer of the demand is not really in a position to command the viewer of the poster. The mood then is one in which the demand that the viewer do something needs to be tempered and perhaps be delivered more as a request than a command. While the name of the event is important information and therefore likely to be presented in a larger scale for emphasis, the greater emphasis on the offering speech act simultaneously affords a more tactful mood. Finally, as I have already alluded to it, typography contributes to the overall coherence of a text. The choice of scale is one way in which salience can be expressed. In the poster, the most important information appears in the largest typeface and the least important appears in a smaller scaled typeface. The choice of typeface can also be used to realize information values and to separate blocks of information. In the poster, both the demand and the offer appear in the jaggie typeface while the lesser details at the bottom appear in a sans-serif typeface.

Layout

Layout serves as another important semiotic resource for the generation of textual cohesion in multimodal texts. Layout, like typography and color, "relates the representational and interactive meanings of the image to each other" and does so "through three interrelated systems": informational value, salience, and framing (Kress and Van Leeuwen 2006: 177). And much like distance and angles of interaction, layout draws upon spatial metaphors to organize compositional elements into meaningful relations. Together, the three systems produce a kind of compositional balance based upon placement, weight, and connections such that a change in the positioning and grouping of elements can effect a change to the entire meaning of the composition (Machin 2011: 129).

Information value refers to the relative significance attributed to the elements of a composition through placement. Degrees of informational value are therefore assigned to the various regions (left or right, top or bottom, and center or margin) of the text and in turn give the viewer cues as to the import or significance of the individual element based upon its placement. Kress and van Leeuwen (2006) propose that left-right compositional relations largely determine "Given-New" information structures while top-bottom relationships tend to determine "Ideal-Real" information structures. In those cultures that use Roman script, the left side has been associated with given and the right with new. This can be seen quite clearly in the way tonal panels are used to organize elements within Figure 3.12 such that the left and right panels could be said to produce both Given and New [the virtually trained soldier (left)—the training products offered (right)] and Idea-Real relations [the well-trained soldier (top)—textual elaboration (bottom left) and corporate logo (bottom right)]. The third type of information structure is, in turn, realized through center-margin relationships. Those elements that occupy a central position within an image tend also to be central to the composition as a whole. Generalizing somewhat, centrally placed components relate to but also hold together the marginal elements of the visual composition. For example, in magazine advertising, it is not uncommon to position the product at the very center of the advertisement and place supporting text at the top and bottom of the ad. Multimodal texts do not necessarily always employ all three information structures, but in contemporary Western forms of visualization, the given-new and ideal-real structures are regularly and reliably utilized. Furthermore, in instances where all three are used, they tend to function together such that marginal elements are endowed with given-new and ideal-real values (Kress and van Leeuwen 2006: 194–200).

Salience is realized through the hierarchical positioning and presentation of elements or "volumes" within a composition in terms of prominence (Kress & van Leeuwen 2006: 66). Those elements that are positioned so as to be more "attention-grabbing" will, accordingly, be treated as more important or of greater visual weight. Salience, Kress and van Leeuwen (2006: 202) suggest, "is not objectively measurable but results from complex interaction, a complex trading-off relationship between a number of factors: size, sharpness of focus, tonal contrast, colour contrasts, placement in the visual field, perspective" and other "quite specific cultural factors." In this way, salience or visual weight is not strictly a spatial feature of compositional meaning but, instead, suggests that there is a balance between elements that the viewer must be able to judge. In the case of the virtual training advertisement, Figure 3.12, while the soldier does not occupy a central position, since the silhouette shape is the largest element of the ad space and we are also predisposed to looking at human forms, it does carry the greatest visual weight. Thus, in judging the balance between elements, the viewer will tend to find what Arnheim (1982) terms the "power of the centre" even when there is no literal organizing center of the image-text (Kress and van Leeuwen 2006: 202).

Finally, framing, the third system in compositional meaning, calls our attention to the ways in which the individual elements of visual compositions are presented in varying degrees of connectedness to the other components. The stronger the framing, the more "boxed in" and, therefore, isolated or disconnected the element is. The result is that the element can be treated as a separate and distinct element within the composition. The greater the degree of connectedness between elements, the weaker the framing, and so the more they will function together as one informational unit. Kress and van Leeuwen (2006: 204) identify a number of ways in which the connectedness of elements can be realized, including the use of vectors and the use of visually rhyming abstract graphic elements, so as to direct the

Figure 3.12 *Use of strong framing in a magazine advertisement.*
Science Applications International Corporation (SAIC) (2006), *Training and Simulation Journal*, vol. 7, no. 1: 11.

viewer's attention from the most salient element to lesser elements within the composition. Thus Figure 3.12 presents an interesting example of strong framing whereby the panels segregate informational units but the color palette actually works to reduce the framing, establishing a kind of overall cohesion of the elements.

Clearly, it is in the composition of the visual and verbal elements that we find the greatest variance between ad images despite their obvious commonalities in terms of ideational and interpersonal meanings. This should not be surprising since the differences do not necessarily mean differences in the thematic meaning of the images but rather that there are always multiple paths to the same meaning formations. At the same time, the ads all do share in certain conventions of magazine advertisements such as the use of a top-bottom structure coinciding with a pictorial "ideal" promise of the product and the "real" of the product.

Conclusion

CDS seeks to not simply describe the representations of society but intervene in those representations so as to transform society. The study of discourse should not be simply a quite scholarly activity that takes place in a modest academic office on a well-groomed campus somewhere. Accordingly, researchers who do CDS seek to take on an advocacy role, not only challenging structures of social domination as they are encountered but actually championing those groups who are subjected to those structures of domination.

I have argued that the framing of the political project of CDS as one of demystifying power relations that otherwise remain hidden to social actors raises ethical-epistemological and, perhaps, even ontological problems for those doing the work of CDS. Instead of conceptualizing CDS as a practice of revealing what is otherwise hidden to social actors, I believe the term, "de-naturalizing" (Machin and Mayr 2012: 5), comes much closer to how we should frame the critical aspirations of CDS. My preference is for de-naturalizing because, to me, it suggests that something has been done with discourse that is not simply a concealment or a repression of reality. De-naturalizing suggests, instead, that something has been made to be experienced as natural and this is closer to Foucault's conception of the positivity of discourse and the productivity of power. At the same time, as Wodak and Meyer remind us, in doing CDS, we should not privilege ourselves as somehow knowing better than those with whom we would side. The politics of CDS is premised upon advocacy and not voluntarism.

The analytic practice of CDS then starts by "making visible the interconnectedness of things" (Wodak and Meyer 2009: 7) and this is done through making systemic descriptions of how things (subjects, practices, facts, etc.) are made to be visible. This, I believe, requires that we adopt the toolkit approach proposed by Machin (2011). To facilitate this, I have outlined some of the principal resources that will be used to develop an MCDA approach to technological discourse. In the remainder of this book, these "tools" will be taken up so as to operationalize the matters raised in Chapter 1.

4

Discourses of technology as progress

In *The Ethics of Rhetoric*, Richard Weaver describes progress as what he calls a "god term." Weaver (1953: 212) uses god term to characterize an "expression about which all other expressions are ranked as subordinate and serving dominations and powers." In many ways, a god term can be likened to a discursive formation (Foucault 1976) or metanarrative (Lyotard et al. 1984). As a god term, progress is a secularized teleology and Weaver describes it as "the ultimate generator of force flowing down through many links of ancillary terms" (Weaver 1953: 212). The prescriptive "power" of a god term is therefore evidenced in its "capacity to demand sacrifice" (Weaver 1953: 214). In his work published in 1953, Weaver proposes that the status of progress is such that it can be used to justify whatever one is able to successfully associate it with. Progress takes on religious connotations so that it "becomes the salvation man is placed on earth to work out; and just as there can be no achievement more important than salvation, so there can be no activity more justified in enlisting our sympathy and support than 'progress'" (Weaver 1953: 213).

This idea that progress is something for which one must make sacrifices is captured perfectly by David Noble when he points out that while some displaced workers might have had their doubts about the benevolence of progress, many still felt the need "to believe that their own sacrifices are suffered for a greater good—how else to suffer them with dignity?" (Noble 1983: 23). Those that must fall by the wayside, left out of the greater good, in order that progress should continue are its unfortunate sacrifices but they are obviously also flawed in some way and not capable of progressing.

This chapter will explore the progress narrative as it was tied to technological innovation. The chapter starts by examining how progress was constituted as a measure of and a means to moral and social betterment. As technology also became a measure of moral and social superiority, it became attached to the progress narrative. Initially part of grand narratives of nationalism in the nineteenth century, progress came to be increasingly more mundane and something to be experienced in people's everyday lives, for better or worse. This is exemplified in the analysis of the Carousel of Progress attraction housed at Disney World in Orlando, Florida. Although progress may not have the ability to compel sacrifice as it once did, the concept of progress still is part of people's everyday vernacular. This is then demonstrated in the analysis of a commercial promoting Honda using its Asimo robot.

Progress as betterment

Progress, as a nominalization, is typically understood as the advancement or forward movement toward a goal or higher level. How the "goal" of progress is defined is, of course, deeply ideological. In his *History of the Idea of Progress*, Nisbet (1980: 4–5) contends that "the idea of progress holds that mankind has advanced in the past—from some aboriginal condition of primitiveness, barbarism, or even nullity—is now advancing, and will continue to advance through the foreseeable future." Progress, understood as a movement toward increasingly superior standards of living and moral betterment comes to be made visible by the technological conditions under which one lives. Not surprisingly then, there is a deeply rooted connection between technology and progress that is bound up in Western colonial and imperialist expansionary discourses. For example, a review of gazetteer entries from the nineteenth century quickly confirms the way popular imperialism legitimated racism and nationalism with such a conception of progress:

> The Different Nations of the earth, are usually divided into the savage, half-civilized, and the civilized. In the savage state, men subsist chiefly by hunting, fishing, and the spontaneous productions of the earth. The civilized and enlightened Christian nations are distinguished for their advancement in science, literature and the arts. (Hodgins 1858: 4)

> The native population of Australia are perhaps the most degraded members of the human family, and the settlement of Europeans upon their shores has done nothing to elevate their social or moral condition. They are few in number, and are fast becoming still further diminished. The colonial population (composed chiefly of British settlers and their descendants) already numbers upwards of a million, and is rapidly increasing. (Maunders 1859: 462)

The first text is typical in that it divides the world between savagery and civilization, but also interesting in that it allows for class of peoples that fall in between. Savage peoples engage in a subsistence existence as "men" while civilized peoples live as "enlightened Christian nations." So while in the first sentence, the earth is divided into different kinds of nations, in the second sentence, "men" are opposed to "nations," thus suggesting that sophisticated forms of socio-political existence, and therefore the goals of progress, are restricted to those living within the "Christian nations." This information is presented or offered in an unprojected or unreported form and so is to be understood as simply fact (Halliday 1985). Written in a highly impersonal form, without an interactant, there is no one with whom to contest the veracity of the claim. After all, being distinguished "for their advancement in science, literature and the arts," would imply a comparative evaluation, but no actual evaluator is represented. Of course, this "objective" truth about the superiority of the Christian nations is ultimately a kind of congratulatory self-address. Likewise, the second text repeats the motif of "the human family," placing the indigenous people at the nadir of progress, and draws upon the trope of the vanishing indigene to euphemistically refer to colonial genocide.

What textbooks such as these did was contribute to the production of a proto-anthropological taxonomy called "the family of man," which legitimated the domination of European imperial nation-states by placing them at the head of the family and therefore as the pinnacle of progress. By comparatively "demonstrating" their moral and material superiority as the result of European progress and enlightenment, the "natural" superiority of (Northern) European cultures was made self-evident; and with it, technology, accordingly, comes to serve as a yardstick for measuring where groups of people fall on "the ladder of predictable progress" (Gould 1989: 35).

In time, progress (social and technological) became more explicitly linked with evolution but it also retained the earlier moral coding. This is largely because popular understandings of evolution are closer to Social Darwinism than they are to the evolutionary theory inspired by Charles Darwin. Social Darwinism is an umbrella term used to refer to a grouping of theories about society that justified the gross inequities of the late Victorian period and the early twentieth century. Probably most closely associated with Social Darwinism is nineteenth-century sociologist Herbert Spencer who coined the phrase "the survival of the fittest" (not Darwin) to characterize a notion of a competitive society in which those who monopolized its resources did so justly out of a process of natural selection. This biological conception of progress, coupled with the earlier moral economy of progress, leads to our thinking about progress as a kind of natural, inevitable, linear process of development toward a perfected state.

Figure 4.1 *Cropped image from "The March of Progress" illustration by Rudolph Zallinger.*

Figure 4.1 is a cropped version of the illustration by Rudolph Zallinger, oft referred to as the "March of Progress," which was commissioned by Time-Life Books for its *Early Man, Volume 1* (1965). Though the original illustration was much longer and encompassed more hominids, this portion should be familiar to most readers. It has likely become such a recognized and mimicked image precisely because it seems to so concisely visualize this linear conception of progress. Though not explicitly referring to Zallinger's illustration, evolutionary biologist Stephen Jay Gould (Gould 1989: 31) declares, reminiscent of Berger on the visual culture of advertising (Berger 1972), that the concept of the "march of progress is the canonical representation of evolution—the one picture immediately grasped and viscerally understood by all. This may best be appreciated by its prominent use in humor and in advertising. These professions provide our best test of public perceptions. Jokes and ads must click in the fleeting second that our attention grants them." The currency of this image is such that intertextual references to it continue to abound. One such image that has circulated widely over the Internet is entitled "The Evolution of Man" and starts with a bipedal simian, and progresses through an empty-handed hominid, to one fully upright carrying a spear, to a homo sapiens slightly leaning forward carrying a rake, to another more stooped carrying a jack-hammer, till finally it evolves into a man hunched over his computer. The joke is obviously that the progression in the kinds of tools we use and the labor we perform has made us, in posture anyway, devolve to some extent. Similarly, Figure 4.2 shows Honda's Asimo robot standing in front of the earlier versions of the robot. They also stand in a single line as if to suggest that each version has evolved as a direct improvement upon the previous iteration. What we can claim, then, is that the image of linear progression has become a resource for representing "evolution" but it affords the same limited understanding of technological development that the march of progress affords human evolution. Just as progress is understood as a

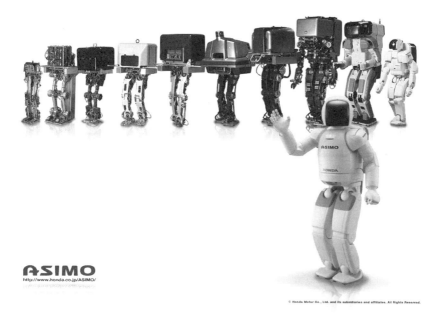

Figure 4.2 *Wallpaper image of various iterations of Honda's Asimo robot. Courtesy of Honda.*

unidirectional movement "forward," technology is also thought of as moving forward, taking its users with it, and bringing them to a better world.

American progress and the vector of improvement

The painting "American Progress" or, alternatively, "Spirit of the Frontier" (1872) by John Gast is often presented as the quintessential representation of the American belief in Manifest Destiny (Greenberg 2005). It was commissioned by George Crofutt and reproduced as a chromolithograph poster and as an illustration for his subscription-based series of travel guides, *Crofutt's Western World*, to promote tourism to the American West.

The painting depicts a woman in a diaphanous white dress floating over what is to become the continental United States, westward of the Mississippi River. She carries a "Common School" book and a coil of telegraph cable, which she is unraveling. In keeping with cartographic conventions, the east is to the left of the painting while the west lies to the right-hand side. The left-hand portion of the painting is fully illuminated while the far right is still cloaked in darkness. In the top left, steamboats and frigates ply the waters and a large suspension bridge spans a wide river. A city can be seen on the peninsula and the far right bank below the bridge. Green fields and presumably pastures extend to the right, but the land becomes increasingly barren and brown to the far right. To the left of the woman, three steam trains move westward parallel to one another. Running alongside the bottom railroad tracks are the telegraph poles that carry the line that she is uncoiling. Behind her foot, a team of horses pull a stagecoach and in front of her a team of oxen pull a covered wagon. On either side of the wagon lie the skeletons of buffaloes. At the bottom right of the painting is a fenced farmstead with a log cabin, and two men are plowing a field with the aid of oxen. Also at the bottom, just to the left, are men who are also moving westward. One is on horseback while the others walk. The man on point carries a rifle while the man at the rear carries digging tools. In the darker right-hand region of the painting, a group of six Native Americans are also moving westward. A woman, with her head down and carrying a child, is being pulled on a travois. The three running figures are all looking backward, presumably at the white people coming from the East and the man riding the other horse sits with his shoulders forward. One of

Figure 4.3 *"American Progress" by John Gast.*

the running figures is a woman who appears to be bare-breasted. Below the Native Americans is a bear, also looking over its shoulder, and above them, a herd of buffaloes. All of these figures depicted in the painting are moving in a common trajectory from right to left.

The painting is clearly intended to be "allegorical," so although there is a kind of action taking place—the migration westward—it would not be accurate to label this a narrative process image (Kress and van Leeuwen 2006). Instead, the painting can be said to be representing a visual conceptual process since the figures on the painting function as carriers of meaning or connotation rather than as specific social actors. In conceptual processes, carriers function in one of three ways: classificational, analytical, or symbolic. Classificational processes represent the participants in some form of taxonomic relationship to one another. Analytical processes represent participants as either the carriers of a specific attribution (the whole) or the parts that are suggestive of the attribution(s) assigned to the carrier. Finally, symbolic processes represent the participants as being representative of or standing for something else. It is this last type of process that best describes how the elements in "American Progress" function.

Van Leeuwen and Kress (2006) further divide symbolic processes into Symbolic Attributive and Symbolic Suggestive image processes. Symbolic Attributive processes include more than one participant, with the meaning of the one participant being realized in relation to the other(s). The meaning or identity of the principal participant or carrier is determined in relation to those of the other participants, which function as attributes. In this type of structure, meaning must be attributed to or conferred upon the carrier. Symbolic Suggestive structures, conversely, entail only one participant that functions as the carrier of a specific meaning embodied in the participant itself. Here the meaning is a property of the carrier itself rather than the product of a relationship between a carrier and its attributes. Symbolic processes can, therefore, be distinguished from analytical ones because the emphasis is always upon the more generalized essential qualities assigned to the participant at the expense of any specific details. In the case of "American Progress," the painting has elements of both Symbolic Attributive and Symbolic Suggestive images.

Symbolic suggestive images differ from attributive ones in that the meaning or identity of the carrier comes from within rather than being conferred upon it. Symbolically suggestive participants, rather than referencing a specific actor, bear a generic quality and connote a generalized essence (Kress and van Leeuwen 2006: 106). Portrayed like a goddess, the floating woman is frequently referred to as Columbia, the Latin-named goddess invented to personify the United States as an imperial power, much like Britannia (Great Britain), Caledonia (Scotland), Germania (Germany), and Hispania (Spain). With the imperial star adorning her head and the glow that she casts upon the land, she is to symbolize the essence of the American (US) empire. Drawing upon European discourses on the nation-state, Columbia symbolizes the enlightening influence of the United States over the continent of North America. As such, she functions as a symbol unto herself and is a semiotic resource for expressing the essence of the United States. Accordingly, as Columbia, the woman in the painting is the spirit that comes to transform the land into the US nation-state.

It is, however, in relation to the other elements of the picture that she becomes more directly tied to progress. While on her own, the woman can be interpreted as a depiction of Columbia, the gendered embodiment of the spirit of the US empire, it is in combination with the presence of the other represented participants in the painting that she comes to function as the symbolization of progress. The steamboats, steam trains, telegraph line, stagecoach, settlers, covered wagon, and even the fleeing Native Americans all function as attributes of a particular discourse on progress in the Americas. These westward-moving transport and communication technologies are attributes of progress as are the fleeing Native Americans and wildlife. The settlement of the West begins with cruder forms of transportation such as the covered wagon and stagecoach and becomes more refined and industrial as they are followed by the railroads and their telegraph lines. In the face of progress, the indigene must flee and so it is not surprising that the representation of the Native Americans looks like a precursor to the "Vanishing Indian" trope made popular in the early-twentieth-century artwork of Frederic Remington and others. However, the woman is attributed to symbolize progress not only through the other figures on the landscape but also by the two objects she carries: the Common School book and the telegraph line. With these objects she brings literacy and enlightenment and joins the people of the continent in a common bond through communication. Together, these attributes of progress, in relation to her connotation of Columbia, contribute to her symbolization of progress.

Progress, then, is a bestowal or a blessing. It is also an expression of the essence of the US nation. Justifying expansionary ambitions and the genocide of indigenous peoples, the painting portrays the forward rush of progress into the dark, "empty" space of the West as divinely sanctioned. Progress in this sense is moral since it is something that is a sign of being among the elect. Progress, therefore, has taken on qualities of salvation, but it is in this life on earth rather than in an afterlife in heaven. So while Columbia may bear connotations of providence, progress, as it is to be actually experienced, is not only moral, but also technological and more mundane than paradise.

The motif of "the taming of the Wild West" is often discursively constituted as a process of domestication. Greenberg (2005: 1–2) posits that the choice of gendering progress as feminine "when so many of the iconic nineteenth-century images of western settlement were male" reflects how the representation of progress and settlement is realized through a process of "benevolent domestication." Gast's painting conceptualizes the disappearance of the wilderness, along with the "wild" flora, fauna, and people that occupied it through successive waves of more intensified settlement. So progress brings enlightenment to the previously darkened regions of the North American continent in the form of white settlement and heralds the arrival of agriculture and industry.

The interactional meanings realized by "American Progress" constitute relationships between viewers and represented participants, between viewers and Corfutt, and to some degree between represented participants. Because the painting is representationally more conceptual than narrative, the only represented participants that interact are the Native Americans and the bear. Though they do not seem to be looking at any of the settler participants in particular but they look eastward and flee their encroachment. All the other represented participants, including the woman as progress, appear to be fixed on moving westward and that is all. In terms of how the painting semiotizes the interaction between Crofutt and his readers, the painting presents a triumphant and self-congratulatory depiction of the "settlement" of the West. Both Crofutt and Gast clearly imagine the audience of the painting to be like-minded in their attitudes toward American progress and to readily understand the chosen iconography. In terms of the viewer's relationship to the represented participants, it is likely one of positive regard (excepting the Native Americans), but also of distance. The figures are all viewed at the side moving westward and do not face the viewer, so there is no "contact" between them. This has the meaning potential of being detached and viewed objectively. This detachment and objectivity is further augmented by the scale of the image. With the exception of Columbia, all of the participants are presented as physically distant figures that in turn have the meaning potential of being socially distant as well. The dimensions of the chromolithograph were only 37.6 × 49 cm, so while there is a great deal of detail, aside from Columbia, the figures traversing the landscape would have appeared quite small. However, the full-length depiction of Columbia also affords an impression of impersonal social distance. Interestingly, the painting adopts an isometric perspective such that the viewer looks upward toward the northern part of the continental United States but at the same time looks down upon the figures crossing the continent. At the same time, the positioning of Columbia places her seemingly at a horizontal angle above the eye-level of the viewer, which has the effect of placing her in high regard. The overall effect is that the viewer is observing concepts rather than interacting semiotically with people and those concepts directly associated with progress, expansion, and domestication are to be venerated.

As a composition, "American Progress" draws upon conventionalized meaning potentials afforded to the spatial organization of information values. The East-West division of the painting results in a left-right division. Van Leeuwen and Kress (2006: 181) argue that left-right structures tend to be homologous with the sequential information structures realized in language. Accordingly, in languages in which writing is from left to right, the horizontal structures in visual composition tend to reproduce given-new relations from left to right. Considering that the subject matter of "American Progress" is the migration westward, it might seem reasonable to assume that the West should be considered new and the East given. However, that would mean treating Given and New as semantic categories rather than functional ones. In this case, the West functions as Given and the East functions as new. Consequently, the information structure realizes the message: "Given the wild, tamed spaces of the West, [New] we must send in people and technology to domesticate it." The very presence and state of the West justifies expansion into it.

The painting can also be said to be structured along the vertical axis with constituent elements occupying the top and bottom of the painting as well as the middle. To the top there is the horizon with clouds and the sun rising to the east. At the bottom there is a stretch of land in which the bear flees the oncoming frontiersmen and a homestead with land turned over to agriculture lies to the East. The top, an image of the heavens, occupies the part of the picture that is presented as the Ideal while the bottom, an image of the land being settled, is positioned as the Real. For van Leeuwen and Kress (2006: 186), the Ideal is "presented as the idealized essence of the information" while the Real represents the "more 'down-to-earth' . . . or more practical information." The painting then presents the heavens as the idealized essence of US progress and expansion while the business on the ground of domestication is the Real. The middle space mediates between the two, depicting the actual sweep of progress across the continent.

Finally, the painting also employs a Center-Margin structure that textualizes the Symbolic Attributive process. The figure of Columbia as progress is centered in the painting and most salient. Positioned as she is, her meaning is constituted through a center-margin relation rather than simply being a function of being in the central part of a visual composition. As van Leewuen and Kress (2006: 196) explain, "to be presented as Centre means that it is the nucleus of the information to which all the other elements are in some sense subservient." As the subject of the painting, the other constituent elements are marginal and therefore subservient to Columbia but also function to attribute the meaning of progress to her.

"American Progress" represents progress through the lens of Enlightenment and imperial values. Progress is tied to the westward expansion of capital and the US state through the settlement of people and the introduction of transportation and communication infrastructure. The painting constitutes progress as moral and technological, but also allows the viewer to experience progress in a linear fashion. The movement westward is accomplished in successive waves, first marked by foot, horseback, and covered wagon, then by stagecoach, and finally by telegraph and train. Progress is therefore represented as a singular trajectory accomplished through successive improvements in technology. Technology is represented in terms of transportation and communication, which is in keeping with the subject matter of the painting. While "American Progress" represents the expansion into the west as a process of civilizing and domesticating the "savage," dark spaces of the continent, the technologies represented are limited to the public sphere and do not include those of the private domestic sphere. Progress is directly tied to the projects of state-building and capitalist expansion. Technology materializes progress, therefore, in the form of privately owned grand projects that bring civilization from the east. "American Progress" therefore articulates a discourse of technology as progress to claim the rightfulness of the US expansion westward and in doing so associates technological progress with domestication as a civilizing project, but not yet the microcosm of the domestic itself.

Textually, then, "American Progress" realizes through vectors of movement from right to left, a line of progressive "improvements" organized under the figure of Columbia, leading to the eventual creation of a common time-space of the "American" nation-state. Representationally, this linear-evolutionary perception of progress supports the claim that the wilds and wastes of the western part of the North American continent await to be domesticated and industrialized by the United States. Like in the illustration by Zallinger, the representation of a line of movement and change serves as a semiotic resource for representing progress as a linear phenomenon. Inheriting earlier conceptions of progress as moral and material betterment, the representation of progress as a linear movement from lesser to greater states of being affords hierarchical representations of progress in which those peoples who are deemed to reside in lesser states of progress can easily be regarded as being of less worth than those who position and privilege themselves at the apex of progress.

Bringing progress into the home

With the Cold War, this connection of progress to moral and material betterment comes to be connected to consumer culture such that the abundance of consumer goods comes to be evidence of the moral, political, and economic superiority of capitalism. This rearticulation of progress from an imperialist mythology to a Cold War one was very much a part of the original Carousel of Progress that was conceived at least in part by Walt Disney. It exemplifies the way in which progress came to be redefined in terms of the democratization of consumption that was intrinsic to Fordism.

Walt Disney's Carousel of Progress is an audio-animatronic attraction that was originally designed and built by The Walt Disney Company for General Electric at the 1964 New York World's Fair. The Carousel was relocated to Disneyland after the closing of the fair and then relocated again to the Tomorrowland section of the Magic Kingdom in Disney World in 1975. Billed as having "had more stage performances than any other show in the history of American theatre," it is effectively an exhibit that showcases how the domestication of electricity heralded the introduction of consumer appliances to family life in the form of new inventions. The Carousel of Progress is interesting precisely because it is so emblematic of the domestication of progress. Unlike the painting "American Progress," the Carousel does not represent progress as an abstract concept but, instead, seeks to represent it empirically as it is experienced in the mundane setting of the "average" American home. While in the painting, progress domesticates the West, in the Carousel, progress is now domesticated. Accordingly, the representation of progress is now realized through domestic consumption rather than grand and sweeping change.

Exhibitionary spaces and built environments like the Carousel of Progress are communicative artifacts and so can be systemically analyzed as "message-entities" (Kress 2009: 59) or textualizations of representational and interactional meaning. As Stenglin (2009: 273–74) explains, the meaningfulness of space is organized through the three metafunctions that all forms of communication perform:

> The ideational function is concerned with the ways we construe our experience. In space, for instance, the *ideational function* is concerned with naming of different spaces and classifying them according to function, e.g. kitchen, bathroom, and bedroom. Space also fulfils an *interpersonal function*, which is concerned with the relationship between a space and its occupant. Third, space fulfils a *textual function* through the organisation of a series of spaces into a meaningful whole.

Particular attention needs to be paid to describing how three-dimensional spaces are organized as meaningfully usable spaces since it is the compositional elements that materialize the representational and interactional meanings of a given space. Since built environments are planned as fixed spaces that must be put into use and occupied by users, they become meaningful as users move through and enact them. Accordingly, Stenglin (2009: 274) proposes that analyzing three-dimensional spaces necessitates having tools for describing how spaces are organized to "unfold dynamically" as well as those for describing the static organization of built spaces.

Stenglin (2009) applies Halliday's textual metafunction to the interpretation of space to provide a useful set of tools for systematically analyzing the dynamic organization of exhibitionary spaces such as the Carousel of Progress. According to Stenglin (2009: 274), built spaces unfold semantically through the meaning potentials afforded by the semiotic resources of Path-Venue and Prominence as well as Given-New structured information values.

Path-Venue structures function to channel the movement of the occupants through a built environment. A Path is a designed route intended to convey visitors toward destination points or Venues. A Path "consists of those spaces designed to regulate 'people flow'" and so works "to channel, distribute and circulate occupants into a building or parts of a building" (Stenglin 2009: 274). Paths are not simply lines but, instead, represent the movements of people through space as they have been designed into the structuring of a built environment. Paths are not always successful as the impromptu trails of dead grass in parks attest and Paths vary in terms of the degree to which they constrain movement, but the most successful ones are the ones that seem obvious to their users. Additionally, Paths do not simply constrain movement; instead, they are "the built medium along which people move." In other words, Paths are affordances that constrain user movements, but they also function as the conditions under which Path-Venue trajectories can be realized.

Stenglin (2009: 274) uses the term Venue to refer to the point to which a Path leads the visitors. Paths not only distribute visitors, they also deliver them. Stenglin observes that etymologically, the meaning of venue is rooted in the Old French "a coming." Venues, therefore, are points in the exhibit that visitors come to in the unfolding of the built environment. At Disney World, attractions such as Carousel of Progress have staging areas where guests collect and view a "pre-show" while waiting to gain access to the actual attraction. Indeed, more popular attractions at Disney World have been updated to provide more venues built into the cueing area so as to keep guests distracted and entertained during longer waiting periods.

Finally, the ability of a venue to draw visitors through a Path is determined by its Prominence. Very much like visual salience in layout, Prominence is used by Stenglin (2009: 274) to address how visitors' attention is planned in the built environment so as to "*draw* occupants out of one space and *lead them* into another." Built spaces, therefore, are designed as sequences of Paths and Venues with systems of Prominence providing the occupants with cues as to how to make the space unfold. Prominence is realized not just visually but can also be realized through other senses such as smell when the aroma of coffee is used as "a powerful lure for many tired museum visitors" (Stenglin 2009: 275).

The Venue is built as a carousel-style theater with a stationary circular stage around which the audience is made to revolve. Six audience sections of 240 seats rotate around an elevated stage that is divided into six equal sections. The first Venue inside the attraction is the staging area where the audience members enter the attraction, find their seats, and listen to a short pre-show welcome and introduction while the sixth and final Venue is the exiting area. The animatronic show takes place in the remaining four sections or Venues with each Venue housing a scene that represents the same family living in four different periods in the twentieth century. The first act of the show takes place in circa 1900, the second takes place in the 1920s, the third in the 1940s, and the final act is set in the present though when it was last updated in 1993, it was to be the year 2000 as imagined then. Because the Carousel rotates the seated audience around the circular stage, the Path to the next Act or Venue is essentially predetermined—though guests who lose interest do get up and leave through the emergency exits. The Prominence of the next Venue is largely realized through the musical cueing that begins just before the audience begins to roll around to the next Act or Venue. The relationship between Venues can, of course, be understood in terms of Given-New information values in which one Venue or Act in the show sets the conditions for the next.

Essentially, the Carousel was designed to pay tribute to the progress made by the "average" American family in obtaining the good life through their use of electricity and new electrical appliances. One of the instrumental versions of the Sherman Brothers' score for the original exhibit in New York was entitled "Music to Buy Toasters by." In its present form, the Carousel still espouses this message. As explained in the pre-show introduction in the first act (the loading stage): "Although our Carousel family has experienced a few changes over the years, our show still revolves around the same theme; and that's progress."

As part of Tomorrowland, the Carousel is housed in a retro-technofuturist round building reminiscent of the original Progressland of the New York World's Fair. To enter the building, "park guests" must walk up an attached ramped walkway with high modern stainless steel railings. The building is covered with a mural of mechanical gears in keeping with the visual theming of Tomorrowland. Unobtrusive speakers near the entrance of the building play the attraction theme music in a 1920s big band swing style, adding to the Prominence of the building entranceway. Once in the building, visitors are ushered to sit in a regular theater-style seating area.

In the first Venue, the Walt Disney's Carousel of Progress cogwheel-style logo is positioned directly in front of the audience over the stage area. The logo is essentially a large purple and gold circle with bright purple gear teeth ringing around it and the circle is divided into top (purple) and bottom (gold) semi-circles with "Carousel" in gold appearing at the top and "Progress" in gold at the bottom. At the top of the semi-circle, "Walt Disney's" appears in white in the signature-style of Disney above "Carousel" and "of" appears in white in a small ellipse at the center of the circle, the color matching the outer cog ring that frames the logo. The colors separate the words but the individual sections are not actually framed, so they remain connected as the title. The typeface is in an art deco style with tall, thin, and rounded monospaced letters. The typeface combined with the placement of the two eight-letter words, "Carousel" and "Progress" in the two semi-circle sections, the color choice, and the "of" in the centered ellipse makes the cogwheel logo appear well balanced. Behind the logo are heavy green velvet stage curtains with another set of shiny black curtains behind them. The curtains and logo are illuminated by lighting that transitions from green to blue to red. The effect of the curtains and the lighting is an aesthetic reminiscent of mid-twentieth-century consumer shows.

The recorded pre-show introduction and the voice of the father are provided by Jean Parker Shepherd who also narrated the film, *A Christmas Story* (1983). His voice has a moderately deep pitch at the creaky end of the phonation scale and the accent is Midwestern. In the introduction he tells the audience in a "folksy" manner that they "are in for a real treat today." The audience is then told how "Walt loved the idea of progress and he loved the American family" and that he "thought it would be fun to watch the American family go through the twentieth century experiencing all the new wonders as they came." Once the introduction is concluded, banjos begin to play along with the theme song. With the start of the theme song, the carousel begins to turn and the audience is taken on the Path to the next Venue.

The theme song, composed by the Sherman Brothers, is entitled "There's a Great Big Beautiful Tomorrow" and the lyrics can be readily found by a quick search on the Internet. The title is also a line from the song and is sung four times in the three-stanza song. Each stanza comprises four lines and the title appears as lines 1 and 3 in the first and third stanzas. The first and third stanzas are essentially the same, so the song begins and ends with the same enthusiastic anticipation of an immanent future that can only be positive. The effect is one of advancement, and also repetition and circularity. True to the title of the song, the two stanzas also remind the listener that each day is followed by another tomorrow, so a shiny new future is guaranteed in the passing of each day, further underscoring the sense of repetition.

Sandwiched in between the first and third stanzas, the second stanza extols the virtues of innovation as a benefit to all. On the one hand, invention is attributed to "Man" in general since there is no definite article before "Man," but on the other, it is the act of an individual while the rest of us can only marvel. In keeping with "the great man/mind" narrative, the source of innovation is the genius of an intellectually and emotionally inspired and determined individual, sexed as male, working alone. Thus the process of invention, in keeping with the hylomorphic model, is one in which the inventor has an inspiration (form) that must then be brought into existence (matter). The invention is a means to something that you or I can only fantasize about and it takes the inventor to find a way to translate our desires into actuality.

In each of the four acts, the father performs as the host/narrator and addresses the audience as guests into the home. Each act is a distinct Venue since it involves another turn on the Path of the carousel, and the theme music played during the transition gives the impending Venue its Prominence. In the first three acts, the stage is divided in three with a larger central stage and then two stages on the left and right, which are hidden behind scrim curtains until it is time for the other family members to speak and interact with the consumer device in question. The scrim curtains act as a kind of porous framing so that family members in peripheral rooms can be brought into the act. With each act, the same family appears, with the members having aged slightly since the previous act. In these acts, the father always appears seated and is shown enjoying the leisure time that progress has brought him. The appearance of the family members together rather than in vignettes orbiting the father, distinguishes the capstone fourth act. The entire family is assembled in the one large living space celebrating Christmas and the father is standing in the kitchen area unsuccessfully cooking the turkey.

As the opening narration asserts, the show literally revolves around the representation of the "ideal" American family and of progress. The family is white, suburban, with middle-class aspirations, and has one male breadwinner as the head of the household. The audience is not told how the father is employed and we are left to assume that the mother does not work outside of the family home. In the third act, the father tells the audience about his commute to work. Very much like the families in the television sit-coms of the 1950s, the family in this act reproduces a very narrow, patriarchal, and white supremacist conception of what the "normal" American family looks like and aspires to be.

In keeping with the promotional origins of the attraction, progress is experienced by the family as a cavalcade of new consumer goods. With each act, the standard of living of the family is raised by electrical appliances that save labor and allow new forms of leisure in the home. Aside from the introduction, there are actually only five verbal references to progress in the dialogue.

DISCOURSES OF TECHNOLOGY AS PROGRESS

In Act 1, progress is referred to as a benefactor: "Thanks to progress, we have a pump right here in the kitchen." In the remaining four explicit references to progress, the family members actually express some ambivalence toward progress when it seems to bring unexpected inconveniences or trivial change:

Act 1

In response to grandmother listening to a record on a phonograph

Parrot: "Squawk! She keeps that thing going all day long. Squawk! Progress!"

. . .

Father: "I think I'll take one of those new-fangled trolleys down to the drug store soda fountain and meet the boys for a cold sarsaparilla. Oh, ha ha. I'm sorry, I forgot, we're drinking root beer now! Same kind of thing, different name. Well, that's progress for you."

Act 3

Father: "I drive into the city for work all day, and then turn right around and drive all the way back. And the highway is crowded with other rats doing the same thing!"

Mother: "That's what they call progress, dear." (Off stage, in a patronizing tone)

Father: "Ha ha ha ha! I guess she's right. But we do have television. When it works (irritated). Gives you something to do after you get home. I kind of like it, you know?"

. . .

Mother is hanging up the wallpaper and painting the basement while father sits in the kitchen

Father: (Sounding worried) "What happened, Sarah?!"

Mother: "Oh, you and your progress! That paint mixer of yours just sloshed paint across my rump . . . er . . . rumpus room."

These moments in the show obviously provide tension-reducing humor, but such humor can actually function to suspend or deflect critical reflection upon the notion of progress and the implicit assumptions of what constitutes progress and to where we should actually be progressing. Humor also affords the opportunity to deflect reflection upon the unequal consequences of progress within the home. In the first three acts, we see how technological progress has given the father the leisure time to sit and chat in the kitchen while for the mother it merely aids her in her work. It allows her to do more washing, with the added enrollment of her daughter, to sew at night, and to turn the basement into a rumpus room. Much like Cowan's (1983) study of technology and unpaid domestic labor, there is always more work for mother. While there may be seemingly some "progress" made in gender equality by the final act when roles are reversed and the father is now standing and struggling in the kitchen and the mother installs a voice-activated smart home system that causes the father to overcook the bird, such reversals depend upon a tacit understanding of "normative" conceptions of gendered labor in order to be interpreted as humorous.

As an "average" American family, the interactional resources employed in the show tend to realize a relationship of familiarity and conviviality between family and audience. In the first act, when the older daughter reacts to being seen by the audience in her undergarments, the father responds, "Don't worry, Patricia. They're friends." The audience is presumed to readily identify with the family as typical and therefore their experiences are considered equally typical. In each of the first three acts, the father proudly introduces the audience to the new domestic technologies that the family has acquired and therefore the progress that society has made. Of course, since the family is trapped in the past, the father erroneously professes that "life just can't get any better"; however, obviously, the audience knows better. Accordingly, the family can be regarded by the audience as somewhat quaint and naïve. The result is that while the show is a tribute to progress, it is also a show that plays upon nostalgia for an innocence often assumed to be lost to the advances of technological progress.

Compositionally, the design of the Carousel of Progress emphasizes circularity and repetition, which also lends itself to a nostalgic perspective on progress and the family. The Path-Venue structure is, of course, circular and the cogwheel logo echoes this. The theme song is also circular in that the first and third stanzas are essentially the same and the first and third lines of those two stanzas are essentially the same. With the exception of the fourth act, at every turn the basic layout of the house remains the same and the audience faces the father sitting in the evolving kitchen. Likewise, the same family is depicted in each act, aging only slightly despite the passing of over 100 years.

Given the understanding of progress as a forward linear movement to betterment, the circular motif and repetition would seem to be at odds with the theme of the show. However, celebrating progress in a circular form affords a way of thinking about progress and technological change in a manner that is non-threatening while still being understood as transformative. So while progress moves forward, some things remain constant. For one, the ideal American family, while made materially better off by progress, remains unchanged and unsullied by technological progress. Time passes, progress continues to bring about the good life, but the family is constant. Thus, the goodness of the family lies in the intrinsic qualities that make them the typical American family. So progress does not actually change the family; instead, progress changes for the better the conditions under which the family lives. What I want to propose then is that the Carousel, on the one hand, reproduces the notion of progress as tied to a linear movement toward material and moral betterment, but, on the other, tempers it, making it less threatening, by couching it in a nostalgic rear-view gaze. The Carousel lets you celebrate progressing forward by comfortably looking backward.

Progress today

Although the notion of technology as progress that we see in Walt Disney's Carousel may seem rather quaint, if not naïve, this view of technology is not exactly archaic. The Honda commercial "Museum," featuring the Asimo robot, quite explicitly draws upon this discourse of technology as progress in its promotion of Honda technologies. In the ad, Asimo is depicted entering, wandering through, and then departing the German Museum of Technology in Berlin museum. Throughout the commercial, Asimo is presented as marveling at the different historical displays of technology from nineteenth-century heavy industry through to the aerospace section of the museum. Asimo is seen to be looking at the exhibits as evidence of progress while, at the same time, Asimo is also being used as evidence of technological progress being ushered in by Honda.

The commercial opens with a medium close-up shot of a stern-faced bust of presumably a nineteenth-century inventor or industrialist. The next shot is a low angle of Asimo's feet as it walks through an open area in the museum. We then see from a low angle, Asimo walking past a doorway. Next, we see Asimo looking up, but this time from a low angle looking past a ship's propeller. Asimo then walks up to look up at a large flywheel on a steam engine. As Asimo leaves the Navigation exhibition, he examines and climbs the stairs to the Aerospace exhibition. On the next level of the museum, Asimo first stops to look at a bucket placed on the floor to contain dripping water and then looks upward to see the source of the water. A drop of water falls on Asmio's visor causing the robot to shake its head. The commercial then cuts to Asimo looking down into the eyepiece of an antique telescope,

Figure 4.4 *Asimo climbs upward.*

following which we see Asimo walking past a row of vintage television sets including one with a video camera at which Asimo stops to regard itself on the screen. The next shot is of Asimo looking up at a space suit and we see its reflection in the space helmet visor. At this point the robot raises its left hand briefly as if tentatively waving hello to the space suit. Asimo is then seen ascending another staircase and the shot then cuts to it walking into an exhibit of older prop-driven aircraft. As Asimo walks through the exhibit, it raises its arms and waves them slightly as if pretending to be an airplane. The commercial then concludes with two shots of Asimo walking through a skyway and out of the building.

The commercial is clearly representing Asimo as being childlike and full of wonder at what is displayed in the museum. Asimo's interest in the bucket, Asimo's stature and body movements are all suggestive of the behaviors of an inquisitive child. The tentativeness at the foot of the stairs and the waving to the spacesuit that looks like a grown up version of the robot are also the kind of emotion-laden behaviors we might imagine to be performed by a child.

One way in which the impression is realized that the faceless Asimo regards the technologies on display with awe and wonder is through the frequency at which the robot is depicted looking up at something in the museum. In the short 90-second commercial, Asimo is shown 9 times looking upward at the museum artifacts. In contrast, he is shown looking downward only 3 times, and one of those times is to look down into the eyepiece of an antique telescope that is pointed upward. Certainly Asimo's height means that it would need to look up in order to see most of the displays, but this again also adds to the childlike quality of Asimo's engagement with the museum. At the same time, it is also worth noting that despite the robot's small stature, the majority of the shots are either eye-level or upward angle shots. With six upward angle shots and twelve eye-level shots, the use of only three downward angle shots is telling. The meaning potentials of these choices in

DISCOURSES OF TECHNOLOGY AS PROGRESS

Figure 4.5 *Asimo pretends to fly.*

angles are strongly suggestive of a positive orientation to the robot. Again, this is entirely in keeping with the sense of watching a delightful and charming child explore a world that is much larger than "him."

The musical score for the commercial also contributes to this impression that the Asimo robot is experiencing the museum as a child might. The music commences at the 36-second mark, coinciding with the sound of water dripping into a bucket placed on the museum floor. The music is a piano instrumental entitled Ba by Goldmund and the musical semiotic resources (see Machin 2010: 98–113) that are drawn upon further contribute to the generation of feelings of whimsy and wonder. The composition is overall in a high key and the cyclical melody is a minimalist one with simple repetitions. The corresponding phrasing entails slow rises and falls in pitch. The timing redounds with the drips of water into the bucket and so creates a more "naturalistic" harmony. The result is that Asimo's tour through the museum is underscored by a slow, natural, and reflective musical composition that redounds with the robot's representation as a childlike innocent admiring the progress of technology.

Additionally, the narration that begins at the 1:05 mark not only makes clear that the commercial has something to say about technology and progress but also adds to the sense that Asimo is like a child whimsically taking in all that has gone before it. The narration is four lines long and is delivered by a man with what could be called a gentle grandfatherly voice. The pitch is lower and soft, the phonation is closer to creaky, and the rate of speech tends to be slow. However, the actual wording of the narration is written in a rather ingenuous manner:

"Technology, making better, better

Onwards, upwards, any way but backwards

Tapping progress on the shoulder and saying,

'More forwards, please!'"

The first line conveys the idea of progress being intrinsically about improvement since better can only get even better, but also, by expressing it as "making better, better," the line takes on a level of unsophistication that you would not normally expect in a commercial about "high technology." It is as if a grandfather were speaking as/for his grandchild, and this is further emphasized by the narrator's projection of "Tapping progress on the shoulder and saying 'More forwards, please!'"

In addition to representing Asimo as an inquisitive child, the commercial also reproduces the linear conception of progress. The second line of the narration obviously does this by commanding technology to move onward and upward. Likewise, the request for "more forwards," like a request for more piggy-back rides, further emphasizes the idea that progress only moves us forward and is a rich source of plenitude and satisfaction. But this conception is also visually reproduced in the commercial. When the narrator says "Onwards, upwards," Asimo is climbing the third set of stairs. At "any way but backwards," Asimo is walking through the airplane exhibit. Additionally, the upward-forward trajectory of progress is also echoed in the movement of Asimo as well as the museum itself. Asimo is only shown ascending the stairs and never descending them in the commercial despite its leaving the building, implying that Asimo, as progress, cannot be seen to move backward. Finally, the second-last shot of Asimo as it leaves the museum by walking across a skyway is taken from a very high angle, which has the meaning potential to raise the status of the robot in the eyes of the viewer.

What the commercial demonstrates is the continued relevance of the progress discourse, albeit more modest in its claims. Progress may not herald the fulfillment of national destiny as it once did, but it still informs people's assumptions about the linearity of technological innovation. As we saw in Figure 4.2, the idea that technologies change through successive improvements is still a very powerful one. In the commercial, especially when compared to Gast's painting, progress is presented as more gentle now. In the Honda commercial, at least, there are no losers or cast-offs. However, the degree to which audiences would accept such a conceptualization of progress as we see in "American Progress" is open to question. It may well be that the non-threatening, childlike naiveté that the robot is made to exhibit makes references to progress more palatable. Ironically, in a time of increasing fears of technological unemployment, we see an articulation of the discourse of progress that makes no hint at sacrifices to come.

Conclusion

In this chapter, I have sought to make clear how the discourse of progress is realized through linear depictions of innovation and change. The discourse of progress has been used to justify conquest as well as moral and biological hierarchies, promote a particular version of "the good life," and present technological change as inevitable and intrinsically good. Progress always passed its blessing on to some and demanded the sacrifice of others. Progress functioned as a kind of divine, grand plan. Even as a secularized narrative, progress still functioned to determine who would be among the elect. Originally, the effects of progress were to be felt on a grand level as we saw in "American Progress." However, progress increasingly began to touch smaller places and became more mundane. Accordingly, progress increasingly became something one could experience personally.

The Carousel of Progress expresses this perfectly. Progress is measured in new things that can be brought into the family home in order to make life more convenient. Once progress begins to be expressed in more mundane settings, it also begins to lose its aura as a "god term." Progress becomes more banal and can be seen in simple things like the purchase of a new stove or a refrigerator rather than great feats of engineering. The Honda commercial comes closer to expressing the older notion of technological progress, but again, it is really more about the promise of new conveniences than the grand aspirations of the nineteenth century.

The discourse of progress makes technological change come to appear as linear innovation. Each new discovery or invention comes from the determination of a "great man" who manages to wrestle it free and make dreams come true. Progress confuses a lineage for a line and treats technologies as if they are predestined forms that need to just be freed from the marble that encases them. The discourse of technology as progress, therefore, disassociates the technology from the associated milieu, making it seemingly fall to earth as signs of salvation.

5

Discourses of technological determinism

There is an Internet meme that uses a picture of a group of eight young people walking single file along a sidewalk on what might be a campus somewhere and all are looking down at their phones. Superimposed at the top of the photo image is the text, "**WHAT'S THE POINT OF BEING AFRAID OF THE ZOMBIE APOCALYPSE**" and on the bottom, "**WHEN YOU'RE ALREADY A ZOMBIE?**" The notion that technology is turning people into zombies or robbing them of their humanity is a common trope. It depends upon a presupposition that technology in and of itself is the primary source of social change.

Technological determinism attributes to technology a degree of agency that makes it able to act independently upon society, remaking it in its own image. As Carey (1997: 316) observes, technological determinism requires understanding technology through a dramaturgical analogy in which technology is "the central character and actor in our social drama, an end as well as a means. In fact, technology plays the role of the trickster in American culture: at each turn of the historical cycle it appears center stage, in a different guise promising something totally new." Technological determinism literally puts technology on the center stage in the role of lead actor and all other elements are delegated to the chorus. And this is the principal problem with technological determinism; it privileges technologies as being autonomous to the social setting or milieu in which they are employed. In other words, technological determinism disassociates the technical object from its associated milieu and yet presumes that it still has the capacity to act upon that milieu. It guides us to seeing only "what the technology can do" while rendering it a self-contained black box to its users.

This chapter examines the ways in which discourses of technological determinism are realized when discussing the relationship between technology and society. Assumptions regarding the causal power of technology mean that it can be as readily cast in the role of villain as it can be cast as the hero in the drama of social change. Technological determinist discourse attributes to technology the ability to change society for the better or for the worse. Accordingly, examples of the way in which mobile technologies and social media are depicted as undermining social relations are used to demonstrate how technology can be constituted as a source of destructive societal change. The chapter then turns to the marketing of the iPhone and the so-called "disruptive technology" in order to explore how technology can just as readily be represented as a revolutionary source of societal change. In each case, social change is presented as being brought on by the introduction of a technology as if all of a sudden it just appeared causa sui.

Technological determinism and causality

Langdon Winner (1977: 75) notes that the "concept 'determine' in its mundane meaning suggests giving direction to, deciding the course of, establishing definitely, fixing the form of configuration of something." Technological determinism, obviously, attributes such causality to technology. It depends upon our tacit acceptance of two hypotheses: "(1) that the technical base of a society is the fundamental condition affecting all patterns of social existence and (2) that changes in technology are the single most important source of change in society" (Winner 1977: 76). Technologically determinist discourse therefore casts technology as the guiding force, and therefore the root of the explanation of how things have come to be as they are. Endowed with such causal power, technology is ascribed the ability to determine the direction society takes for better or worse, leaving it open to either praise or blame (Slack and Wise 2005: 42).

In order to play the role of determining agent, technology is constituted with three key characteristics or qualities according to Slack and Wise (Slack and Wise 2005: 102):

1 "Technologies are isolatable objects, that is, discrete things"

2 "Technologies are seen as the cause of change in society"

3 "Technologies are autonomous in origin and action"

The first point means that technologies tend to be understood as self-evident and self-contained things. Technologies tend to be conceived of as objects that are useful and the implications of a technology are to be found, straightforwardly, in the inner workings of the device. The second point, as it has already been discussed, means that technologies have the capacity to act upon society and, indeed, are objects that drive change. The third point is directly connected to the first two. Technologies are perceived to spontaneously appear and thereupon make their consequences felt. This is precisely why thinking in terms of technological impact is an ill-conceived approach to understanding technoculture. To think in terms of impact reinforces the belief "that technologies appear as though motivated by some inertial force that exists apart from the goals, motivations, and desires of human beings and apart from the organization of culture" (Slack and Wise 2005: 102–03). As Akrich (1992: 206) so succinctly puts it, "Technological determinism pays no attention to what is brought together, and ultimately replaced, by the structural effects of a network."

Technological determinism redistributes the agency of the technological apparatus by disconnecting the technical object from the interconnections of the network in which it is embedded and treats it as if the lines of force solely emanate from the object itself. The result is that to adopt a discourse of technological determinism means casting ourselves in the role of bystanders or benefactors of either good fortune or ill fortune depending upon the effects of the technology. This also means that technologies, by virtue of being technologies, have to produce effects upon society leading us to imagine that every technology can be a progenitor of societal change. Thus, technological determinism tends to treat technology in a binary fashion as the source of either destructive or revolutionary change.

Technology as destruction

One way in which technological determinism is discursively realized is by characterizing it as a caustic force that acts autonomously upon and undermines social relations. This is largely typical of reactionary responses to new media, which see the emergence of a new technology as a "sudden arrival" that threatens to destroy the established and, therefore, better ways of doing things. The problem does not lie in adopting a critical eye in the face of new technologies but rather in looking at technologies, both old and new, in terms of being more or less authentic. Furthermore, such a viewpoint is an impoverished form of critique since it can only comprehend social phenomenon in a reductive way by attributing all agency to the technology itself. This can be seen in the way in which some critiques of social media and mobile technologies have been formulated so as to assume that the users of the technologies are just hapless victims of the supposed seductive powers of the technology.

The YouTube video, *Look Up*, written and produced by Gary Turk gained viral status in the Spring of 2014 after being spurred on in part by tweets from celebrities such as talkshow host Andy Cohen, Wimbledon champion Andy Murray, and American Idol winner Jordin Sparks. By May 10, it had over 32 million "views" on his YouTube channel (Carbone 2014). The video is a poem exhorting viewers to put away their electronic devices and communicate with one another face to face rather than through social media. As media reports were quick to point out, it was, ironically, a phenomenal social media hit.

In the video, Gary Turk acts as narrator, reciting his poem in a slam poetry style. He appears with a stubble beard and wearing a crew neck sweater with buttoned-up shirt, slightly exposed in front of a black background. His attire affords an impression of casualness and, therefore, familiarity. He always appears in the video looking straight at the camera, framed in a medium close-up (head and shoulders), which evokes a personal social distance in Northern European proxemics. The use of low lighting perhaps ensures that he does not appear as overbearing. Additionally, there are two other principal actors who undergo a transformation by the end of the video and put down their devices. They also play the couple in the love story that takes up about half of the video.

As the video proceeds, we see scenes of people using mobile devices and a laptop to access social media, alone and in the presence of others. There are two scenes (a field overlooking the chalk cliffs at Dover and a living room) in which a woman interacts with her friends but then is revealed to be alone interacting over a phone. The next scene is of a man sitting on the couch engaged with his phone and a barely visible other man sitting off screen at the end of the couch. The next of the vignettes is of the woman sitting at a bus stop and being snubbed by the two seated women who ignore her greeting and continue to text. After another shot of Turk, the next scene opens with a man seated at a table using his phone with half a bottle of wine in front of him. The camera then pans to reveal his female dinner companion who is clearly looking uncomfortable. The poem then shifts to children and parenting and we see a girl looking at a screen wearing headphones, and a small boy on a tablet. At this point there is another contrasting scene in which children are depicted playing in a playground and then the playground appears empty with swings blowing in the wind. The video then returns to kids indoors with one playing a first person shooter and the same small boy using a tablet. Then, at the mid-point of the video, a major sub-plot is played out in which a young man asks a young woman for directions, and in compressed form, we see the "highlights" of their lives together until in old age, the woman passes away. At this point, it is revealed that this never happened because instead of a piece of paper with an address written on it, the young man had a phone and relied upon it instead of the kindness of a stranger. At this point, in the time-lapse photography style of *Koyaanisqatsi* (1982), we see the same young man standing alone on a busy sidewalk thumbing over his phone screen as people quickly move past. Ultimately, the video comes to its conclusion with the woman looking up and at herself in the mirror and the man turning off his iPhone and leaving it at home as he goes out. Turk then appears again to tell the viewer: "Stop watching this video. Live life the real way."' The final shot is in the park again at ground level with a child's trainer running across the grass.

Interactionally, the video is a combination of admonishment and edification. The poem itself is told primarily in the first person singular (I) and plural (we) and in the second person singular. Turk's appearances in the video tend to coincide with the use of first person singular and plural points of view. Turk's recitation is meant to make you feel as if he is addressing you personally, sincerely, and eye to eye. His use of first person is almost entirely in the subjective case and while he does use second person in the objective case (to show consequences) considerably more frequently, the majority of the time it is subjective. In terms of modality, the poem is written with a high degree of certainty, and in keeping with the slam poetry style, he delivers his poem with a strong sense of conviction. Turk does use the third person singular point of view (the pronoun she) when he tells the love story, which suggests he is speaking to the man or rather the viewer who would identify with the man. Turk effectively positions himself as the more experienced and wiser "mate" who has it figured out and is trying to make you see what you stand to lose if you do not "go out into the world" and "leave distractions behind" (Turk 2014: line 86).

The musical score for the video is an instrumental provided by New Desert Blues, which begins to slowly rise at the start of line two of Turk's poem. A piano is played with the same note being repeatedly struck at a fairly quick timing while another guitar is slowly strummed in threes with progressively softer and lower chords. In terms of phrasing then, we have a long decay, which suggests "a lingering in the emotion" (Machin 2010: 112). Playing in this style has the meaning potential of thoughtfulness and reflection. As Turk's delivery increases in speed and volume, the piano rises in pitch complementing the rise in emotion. Ultimately, the effect of the low pitch range of the piano coupled with the limited pitch range and slow tempo of the guitar is one of intensity and consternation. During the park sequence of the video, the piano is played at a much lower pitch with the guitar strums being replaced by lower octave piano notes, which underscores the seriousness or gravity of the situation. At the end of the playground scenes when Turk speaks to the viewer again, a keyboard is introduced with quiet marimba sounds to transition to the romance sub-plot of the poem. At that point, another guitar playing a quicker tempo and more distorted riff is introduced, providing a very different more "positive" sound to separate this part of the poem from the earlier admonishment. This guitar sound begins to fade as it is revealed that the romance never actually happened, and the original score returns.

Representationally, "Look Up" depends upon a neat and straightforward distinction between "the real world" of face-to-face communication and the "artificial" realm of social media. Turk begins by presenting a conundrum: he has many friends but is, nonetheless, lonely. On the one hand, you can interact authentically by looking into someone's eyes and, on the other, you can interact disingenuously with just some name that appears on the screen. This leads Turk to a moment of epiphany—opening a computer is like metaphorically closing a door. It is the artificiality of screen interaction and friendship that fails him. Accordingly, Turk sets up a number of oppositions between the socialness of face-to-face communication and the anti-social qualities of computer-mediated communication (CMC). Turk juxtaposes eye contact with mediated contact (or, more precisely, looking at a screen) three times in the poem, and the romance that never was begins with eye contact. Furthermore, Turk goes on to claim that the encroachment of the inauthentic world of CMCs upon the real world has created "confusion" (Turk 2014: line 12). It is not just that the "digital world" is illusory; it actually has deleterious effects on social interactions in the "real world."

This leads him to present a second conundrum: our mastery of technology has made us slaves. Our use of new media technologies "where we all share our best bits but, leave out the emotion" (Turk 2014: line 16) has resulted in "us living like robots" (Turk 2014: line 38). At another point in the poem, Turk declares that, in contrast, seemingly solitary acts of reading, painting, and exercising are fundamentally different because "You're being productive and present, not reserved and recluse" (Turk 2014: line 27). Turk also comments on generational changes in children's recreation, observing that as a child, much of his time was spent roaming and playing outdoors while today, he says, the parks are empty. This is reinforced throughout the video by setting all of the images of children using new media devices indoors while adults are represented using their phones outdoors. The upshot is that Turk has moved from criticizing the illusory socialness of social media to the users themselves. Accordingly, Turk presents a third conundrum: "smart phones and dumb people" (Turk 2014: line 48).

This supposed inauthentic quality of CMCs is reinforced by the way in which Turk represents technology and social media. In the poem, social media is reduced to "a screen," "a menu," and "a contact list." This use of synecdoche, where a part stands in for the whole, has the effect of further trivializing CMC but stressing that it is not a person that you are interacting with but rather just a small trivial thing. Turk also refers to CMC devices in general as "a device of delusion" and technology as producing the illusion of companionship and inclusion. By contrast, the romance-never-to-be is presented as what a "real connection" would look like. Turk also uses other metaphors for the social media-technology nexus such as a net in which the user gets caught and a machine. Each metaphor, in its own way, diminishes the agency of the user and privileges that of the device, but aside from a brief reference to a "rich greedy bastard" selling information, there is no real consideration of the broader context or milieu in which mobile CMC technologies have come to be used. Life under social media is lonely because of the technology itself and the way in which it supersedes "real connections."

"Look Up," then, can be understood as an example of the way in which technological determinism is discursively articulated to moralize about the present by holding it up against an idyllic past when we had real connections to one another through face-to-face contact, children played with one another in the great outdoors, and parents spent one-on-one time with their children and did not use technological devices as babysitters. The lesson is that technology is ensnaring us, making us forget how to be together in person and if we can put the technology away, we can reclaim this truer prior form of sociability.

The readiness with which we recognize this discourse of technological determinism and apply it to new media is also the basis of the humor in the Coca-Cola commercial, Social Media Guard (2014). In the ad, people are depicted in typical social settings such as walking along the beach, sitting in a restaurant, at the family dinner table, in the family room, and so on. In each case, the people are depicted looking down and interacting with their screens rather than with the people in their immediate proximity. Coca-Cola's fictional "Analog Lab" offers a low-tech solution to the problem of when "social media can get in the way of the real world" (Coke: 2014). The guard is essentially a Coke-branded Elizabethan collar such as you would put on a dog to prevent it licking at a wound. The effect, of course, is to block downward vision and create a horizontal plane at eye-level. Suddenly, people can see those around them and begin to interact with one another. So whereas the serious exhortation of "Look Up" depends upon human will to overcome the determining effects of mobile technologies, "Social Media Guard," instead, seemingly proposes the determining power of another "analog" technology to counter the determining effects of mobile technologies. Of course, in reality, "Social Media Guard" is also making a claim similar to "Look Up" in that through hyperbole, it is, in effect, also exhorting the viewer to "just look up."

There is, of course, a long history of imagining users of a new media to be stupefied. Just as audiences of mass media were characterized as being passive and unwitting dupes, one way in which social media users are perceived is as unwitting and trapped by a seductive new technology. In 1906, *Punch* cartoonist Lewis Bavner produced a cartoon entitled "Forecasts for 1907" that expresses this sentiment quite clearly. In it are depicted a man

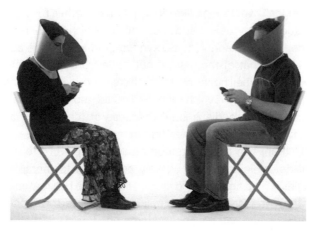

Figure 5.1 *Technology defeats technology.*

and a woman, both of means judging by their attire, sitting in garden chairs at oblique angles to one another. Both have antennas protruding from the top of their heads and wireless teleprinters on their laps. Not only do the subjects not face each other, but both also look down toward the telegraph ribbons that are being printed. The caption below the image elaborates that the two are not communicating with one another and that the woman is "receiving" a romantic message while the man is receiving racing results. So, despite being in close proximity to one another, under a tree in what can be presumed to be a pleasant day in Hyde Park, both people have their heads down and their attention is elsewhere. The bodily orientations of the two people make it clear that they are not interacting. The emotions expressed by each are equally at odds: she is enamored and he is clearly embittered by the race outcome. The explanatory caption underscores that the two are "not communicating with each other" and represents them, instead, as "receiving" messages. In terms of transitivity, they are being passivated since the caption clarifies that they are not engaged in an active action or process (communication) but, rather, are engaged in a more passive one of "receiving," beneficializing from the telegraphy transmissions.

Technology as revolution

Just as technological determinism opens the door to laying blame upon technology for societal ills, perceived or otherwise, it conversely also opens the door to celebrating it with praise. In this case, technology is characterized as a benevolent actor bringing positive change, and usually "opportunity," with its arrival. This is effectively the other side of the same determinist coin as "technology as destruction," but rather than being a weak critique of technology, it is, instead, rooted in a more celebratory discourse. Technology is still understood as being made up of isolatable, autonomous objects that bring about change in society, but the changes they bring are all for the better. In this case, technology is understood as closer to the progress narrative that was addressed in the previous chapter, though as we see in the disruptive technology examples, the two are not quite the same. One key difference is that technological determinism, though it clearly informs the progress narrative, does not presuppose the intrinsic movement toward a perfected state, only that the present good has been ushered in by technological change. In the disruptive technology example, technology does not bring about progress so much as it does opportunity. Likewise, in the iPhone commercial, the technology brings about change for the better, but it is a far more personal and mundane form of change than is constituted in discourses of technological progress.

There'll always be a technical solution

Apple's 2009 iPhone commercial that included the slogan, "There's an app for that," is an exemplary example of technological determinism and what Evgeny Morozov (2014: 28) calls technological solutionism—"recasting all complex social situations either as neatly defined problems with definite computable solutions or as transparent and self-evident processes that can be easily optimized." Intended to promote its app store in tandem with the iPhone, the slogan soon gained meme status on the Internet and Apple applied to trademark the slogan in December of that year. "There's an app for that" expresses precisely the belief that whatever you need to accomplish, there will always be a technological solution and that technology is, in and of itself, what will make life better.

The commercial is 30 seconds in duration, comprising 13 shots. It opens with a hand holding an iPhone in front of a white background. The iPhone is framed so that the top and bottom of the device extend to nearly the top and bottom of the video frame, thus making it the focus of the viewers' attention and, therefore, of principal salience. Throughout the commercial, the shots of the phone change from full to medium close-up to close-up depending upon the action taking place, but the phone always remains centered in the frame. In keeping with much of Apple advertising, the ad is composed with a neutral white background so that the product has little to no visual competition for viewer attention.

Figure 5.2 *Swiping gesture being demonstrated in iPhone commercial.*

The focus of the commercial is to show the iPhone in use not as a phone through which to converse or send text messages but as a personal computing device. The device is represented as a kind of portal on the world in which, by acting on the device, "real world" information is quickly and easily delivered to you. Interacting with the phone is demonstrated as a simple set of gestures that can be done with one finger contacting the touch-sensitive screen: swiping horizontally, tapping, and scrolling vertically. Each gesture causes a desired change on the phone's display (see Table 5.1). Furthermore, by demonstrating the phone in use against a neutral backdrop, no contextual information is offered for its use beyond what is offered in the narrative. The interaction with the iPhone could be taking place anywhere, so its use is not bound to specific settings and circumstances and is, instead, governed by user desire.

The voice-over narrative is delivered by a male who speaks informally and assumes close alignment in opinion with the viewer. The initiating statement begins with "What's great about the iPhone" which presumes that it is already agreed between narrator and viewer that the iPhone is great. The use of contractions on the part of the narrator also has the potential to realize informality and familiarity. The narration itself is made up of a series of statements (giving information) in the form of four propositions. In each of the first three propositions, a scenario is established: "if you wanna" Had the information been presented in interrogative form, as an offering of a good or service, the viewer would have been asked, "do you wanna?" and in which case the viewer's response could have been "no," thus making it pointless to tell the viewer that there is an app available that can do so. Accordingly, the narration proceeds as if the interests and desires of the viewer are closely

Table 5.1 'There's an app for that' commercial by shot

Shot	Type	Gesture	Action	Narration
1	Full	Swipe Swipe	Navigate to app	What's great about the iPhone . . .
2	Close-Up	Tap	Launch app	is that
3	Full	Swipe Swipe	Navigate in app	if you wanna check snow conditions on the mountain,
4	Med Close		Return result	there's an app for that.
5	Full	Swipe	Navigate to app	
6	Close-Up	Tap	Launch app	If you wanna check
7	Med Close	Tap Scroll Tap Scroll	Select input info	how many calories are in your lunch, there's an app for that.
8	Close-Up		Return result	
9	Full	Swipe Swipe	Navigate to app	
10	Med Close	Tap	Launch app	And if you wanna check where exactly you parked
11	Close-Up	Tap	Run function	the car,
12	Med Close	Tap	Input parameter Return result	there's even an app for that. Yep, there's an app for just about
13	Closing	Row of app icons "iPhone 3G"		anything. Only on the iPhone.

aligned with those of the narrator. In short, the commercial promises near unlimited functionality.

Each of the three scenarios entails a desire to know something and the need to consult the "outside world" to obtain the needed information. In the first, it is about snow conditions elsewhere, in the second it is how to get back to the car, and in the third, it is what lunch would translate into as calories. In each case, an app allows the users to interact with a database stored somewhere else giving them extra-local access to information that would otherwise be unavailable. Furthermore, this has the effect of dividing the interests and desires of the individual into a series of distinct microtasks that have corresponding unique programmatic solutions. Life can literally be broken down into apps. Together, the software on the iPhone hardware can enable the individual user to do just about anything her or his heart desires because "there's an app for that."

Visually, the actions of the commercial are a series of deictic gestures enacted by the holder of the iPhone. As we see in Table 5.1, with each scenario, a computational task is performed to demonstrate how the user can discover the particular desired piece of information. In each case, using the iPhone is presented as a matter of find, launch, and discover through a simple sequence of gestures. The series of swipes and taps used to navigate and then launch each of the three apps coincides with changes in the distance of the shot. Each time the user navigates to the app, it is done with using a full shot of the iPhone, but then when the app is to be launched a close-up shot of the phone is used. Once launched, a medium close-up is typically used to show the information being supplied by the app. A full shot is used for the snow conditions, but that is most likely because the app displays the information using a full screen. The use of the phone is, therefore, represented as one in which the user is able to readily find on the screen whatever is wanted through simple indexical gesticulations.

In their social semiotic account of the smartphone, Kress and Adami (2010) offer an analysis based upon hardware and software affordances. The authors note that the increasing emphasis upon screen size and resolution when compared to traditional mobile phones is indicative of an emphasis being placed upon their "'convergence' characteristics" and that "visual output has clearly become the priority" (Kress and Adami 2010: 187). This can be seen in the dominance of the "bar"-type form factor for smartphones in which the screen occupies most of the front surface of the device. Designed in this way, smartphones such as the iPhone are literally built for viewing visual content at the 16:9 aspect ratio. At the level of software, Kress and Adami propose that the inclusion of manipulatable visual screen objects affords an "aesthetics of interactivity." The object-oriented Graphical User Interface (GUI) has become the standardized form of menu on mobile devices and has led "to a conception of semiotic (inter-activity as a matter of 'navigation' and selection among options" (Kress and Adami 2010: 188). The result is that "semiotic action—whether as representation, production or communication—is coming to be seen as selection-driven" (Kress and Adami 2010: 188) rather than, say, "production" oriented. Thus, through the affordances of the software, selection comes to be the principal expression of semiotic agency within the aesthetic of interactivity.

The iPhone commercial realizes this aesthetic of interactivity by focusing upon the iPhone as an object in use. The ability to select is precisely what the commercial is about. This aesthetic of interactivity simultaneously endows the device with the apparent power to do things in "the real world" and the user with a sense of control and mastery *through* the device. The iPhone holder demonstrates for the viewer just how easily information can be selected from the "outside world" by interacting on the screen of the iPhone. With regard to the iPhone, Kress and Adami (2010: 188) note that "there is both *logical and spatial* proximity/continuity of action and effect—a touch on the object moves it—so that the perceived gap between real and handling of objects is narrowed." Thus, the privileging of the visual mode in the design and functionality of the smartphone affords a particular way of interacting with and using the device that realizes a "motion-as-(inter-)activity" (Kress and Adami 2010: 188) metaphor.

To use a smartphone entails interacting with the device through deictic gestures that manipulate objects on the screen, which in turn produces information called upon to represent "the real world." What Kress and Adami are describing when they refer to the relationship between action and effect then is the way the smartphone functions as interface, affording traffic between the symbolic and the material. The interface, according to Hayles (2002: 22), can be understood as a material rather than figurative metaphor insofar as there is a transfer "between a symbol (more properly, a network of symbols) and material apparatus." The iPhone as both software and hardware affords interactions with the world through functionalities that are semiotized on the screen and in the gestures of the user. Accordingly, we can conceive of the iPhone as a material metaphor that both semiotizes and materializes the "power to make things happen in the real world, for it is connected to a complex material apparatus that operates machinery as well as such socio-material constructions as economic transactions" (Hayles 2002: 22). What we see in the ad then is a representation of the iPhone as an interface that allows easy traffic between the information about the world "out there" and the display that occupies much of the face of the device.

Agency on the part of the user, then, can only be realized in the act of selection. In the commercial, every want has a set of corresponding deictic gestures that is rewarded with the delivery of information. The user navigates through layers of options delivered by the technology (choice of which app on the phone, choice of which options within the app, and choice of which apps available for the iPhone), which are then to be experienced as the exercise of agency. Simply put, the iPhone is presented as being like your genie in the bottle. Accordingly, the smartphone as the individuation of a myriad of different technologies (capacitive touch-screen, LED display, battery, wireless and cellular, micro-processor, sensors, digital camera, microphone, etc.) is concretized in relation to an associated milieu that emphasizes individualization, mobility, and connectivity through CMC. Both the iPhone design and its representation in-use materialize "an ideology of choice" that "pervades the social domain" (Kress and Adami 2010: 184). The iPhone is presented not only as the means to exercising choice but also as a platform that makes individual choice and satisfaction possible on the go. Ultimately, the commercial falls back into the narrative of "Look what technology allows us to do."

Technological determinism and technocratic discourse

One of the typical ways in which technological determinism is frequently semiotized is through technocratic discourse. McKenna and Graham (2000: 219) propose that technocratic discourse "has become a recognizable feature of public policy, business, and the social sciences." They argue that technocratic discourse makes claims to objectivity by drawing upon scientific discourse in order to present statements that would otherwise be open to debate as fact. Thus, technocratic discourse bears the appearance of being objective, rational, and incontrovertible while being "precisely the opposite in most cases" (McKenna and Graham 2000: 220).

Elsewhere, Graham (2001) offers three key features of technocratic discourse. First, it is factual in appearance. It rarely contains explicit exhortations since, lacking the aura and authority to command publics as science once did, technocratic convention is "to rationalize their 'proposals' for action with seemingly rational statements of fact, so that they seemingly function as 'propositions'" (Graham 2001: 766). This distinction between proposals and propositions is derived from Halliday (1985: 70–71) who clarifies that propositions entail the exchange of information and are therefore statements that can be contested, while proposals, on the other hand, entail the exchange of goods and services as offers and commands, which can be refused but not actually debated as propositional statements and questions can. What this means is that technocratic discourse presents demands for action in the guise of statements of fact. Secondly, technocratic discourse seeks to rationalize claims to truth, thereby compelling audiences to comply through the use of tense. As Graham (2001: 767) demonstrates, technocratic discourse uses "the tense system to portray the future and imagined states as if they actually existed in the here-and-now." According to Graham, this is frequently accomplished through the use of process metaphors and nominals that are embedded with potentiality such as "opportunities" or "possibilities" or "prospects." Finally, the third feature of technocratic discourse is the presentation of the imagined future "as an extremely *Desirable*, and thus *valuable*, place." Indeed, Graham goes on to characterize the imagined state to which technocratic discourse proposes to lead us as utopian. And, of course, it is seeing the evidence as presented in the technocratic discourse that will ensure our arrival at these utopian states.

Graham's focus is obviously upon policy documents, but technocratic discourse is certainly articulated in other genres as well. The following text is a description of the factory workplace of the future produced for Siemens' promotional magazine. In it we can find some elements of technocratic discourse being drawn upon in order to cast the envisioned workplace as an immanent utopian space.

Coming Soon: Personalized Factory Workstations (Gold 2013)

1 Tomorrow's factory jobs will be completely different from those of today.

2 Although they will continue to be organized around assembly stations, they will not work in rigid shifts, be subject to inflexible processes, or be restricted to a single workstation.

3 According to Johannes Scholz and Johannes Labuttis, engineers who studied production management and ergonomics at the Technical University of Munich, in fifteen years, most monotonous and strenuous activities will probably be a thing of the past.

4 Scholz and Labuttis now work at Siemens Corporate Technology in Munich, where they focus on the role of humans in production processes.

5 "In the future, workers will use their smartphones and computers to organize their shifts themselves," says Scholz.

6 "When doing so, they will be able to take into account their personal chrono-biological attributes—for example, whether they're day people or night people.

7 "This will enable them to adapt their work assignments to their private needs and personal situations."

8 The idea here is to optimally align an employee's individual time management with a company's human resource requirements.

9 This is important, as Scholz points out, because the factory of the future will be highly flexible and organized like a type of living Internet in which everything, and everyone, is networked.

10 "Production lines and their individual assembly stations will be transformable, and it will be easy to retool them in line with the customer order in question," Labuttis explains.

11 This will enable quick reactions to changes in demand.

12 Workers will switch from one assembly station to another at a defined pace.

13 They will be knowledgeable about all the steps involved—from production of individual work pieces to final assembly.

14 Plant managers will benefit from this because networking will allow them to deploy workers in the most efficient manner at all stations.

15 Because everything will be networked, each workstation will "know" at all times which employee is scheduled to work at it next.

16 It will then adjust its parameters accordingly within seconds.

17 Tool placement will be personalized and optimized, and all height- and angle-adjustable equipment will be adjusted to the worker's size, taking any employee limitations or disabilities into account.

18 "The variations will be as individualized as the workers themselves and could include things such as standing aids, footrests, and even a completely different workstation layout," says Labuttis.

19 Robots will also be part of the picture, helping with things such as heavy lifting.

20 Tomorrow's factories will be both productive and flexible, meaning that humans will provide flexibility while robots will ensure fast and efficient production.

21 The average age of factory employees will also change.

22 In particular, workers in today's industrialized countries will be significantly older due to the rapidly progressing demographic transformation that is already under way.

23 By 2050, for example, the number of people over 65 around the world will triple from the current figure of 500 million.

24 People will have to work longer if social security systems are to remain affordable.

25 However, older workers will also be urgently needed because of their skills, knowledge, and experience.

The first thing worth noting in the text is the preponderance of the will+infinitive compound future tense. Only sentences 4 and 8 do not use this form and this is because sentence 4 is an extension providing further information about the two researchers and sentence 8 is an elaboration upon the information provided in sentence 7. The language in this text is being used to make predictions about the future of factory work. It describes how factory work in the future will be organized and distributed, and through the use of future tense it does this as if they are matters of fact.

The second thing to note is the lack of probability modifiers. Halliday (1985: 335) refers to clauses in which information is exchanged as being the indicative type and clauses that function as an exchange of goods and services as the imperative type. Each type of clause can be expressed along a continuum of positive to negative. Indicative type clauses realize degrees of probability and usuality and since the future tense clauses above are, with only one exception, without probability or usuality markers, the futures they predict have the potential to be understood as being near certain. In sentence 3, the only use of low probability modalization occurs where it is stated that "most monotonous and strenuous activities will probably be a thing of the past." Interestingly, the subject here is the kind of work that is being associated with present-day factory work rather than the future of factory work. Whenever the future of factory work is the subject, probability is never modified, which further adds to the impression that the predictive statements are to be read as matters of fact.

The third thing to be noted is the way the future of factory work is opposed to factory work as it is presently organized. In sentence 2 it is stated that in the future, factory work will not be rigid, inflexible, and restricted, which, by implication, is characteristic of present-day factory work. In contrast, sentences 9 and 20 characterize future factory work as being flexible; sentences 6, 7, 17, and 18 describe it as personalized; sentences 5 and 9 describe it as self-organizing; and sentences 9, 14, and 15 describe it as being networked. The factory of the future is thus presented as a neoliberal utopia in which everything is just-in-time and on-demand. Workers come and go as and when needed and are always optimized at peak efficiency. Furthermore, as part of a "living Internet," they have internalized management such that they organize their own shifts. Indeed, the only mention of management at all is in sentence 14, where more senior plant managers are beneficialized.

What this example demonstrates is that future tense and modalization can function as resources for producing accounts of the future that seem factual but also utopian. The factory of the future is in actuality an irrealis space, (Graham 2001) that is a potential space being presented as a reality that is not only predictable but immanent in the current present. It is technological innovation in the form of networked communications and adaptable smart workspaces that ushers in the utopian factory of the future where everyone's needs can be met while maintaining a workforce that must always be on-demand.

Another example of technological determinism that attributes technology with generative rather than purely destructive agency is the concept of disruptive technology. The concept of disruptive technologies is introduced by Christensen (Bower and Christensen 1995; Christensen 1997) to describe a process of destabilization brought on by technological innovation that leads to established companies being displaced by new rival companies. The introduction of a new technology brings about profound changes in the marketplace as companies that previously dominated a market are undercut in price by newcomers offering simpler or more convenient alternatives to their more expensive goods or services. In this case, technology is thought to be the cause of change, but this time it is creative rather than destructive. Not surprisingly, Christensen's concept of disruption has been compared to Schumpeter's (1950: 139) theory of creative destruction though it overlooks Schumpeter's claim that the process is not sustainable and will ultimately lead to capitalism hollowing out its own institutions. As a theory of innovation, disruption follows a thermodynamic model in which the introduction of a new technology forces the system to arrive at a new equilibrium. Like all forms of technological determinism, it treats technology as essentially autonomous such that it seems to spontaneously come into existence from outside the system and then impacts upon that system.

The Deloitte Australia promotional video *Digital Disruption—Short fuse, big bang?* (2012) exemplifies this treatment of technological innovation. The video opens with the ignition of a sparkle or fuse flame from which letters cascade out. The narrator states, "Disruption, it used to happen occasionally," and then a line appears across the frame to become a burning fuse. The line then changes again to become a heart trace as you would see on an electrocardiograph. This has the potential to evoke the concept of technological disruption quite effectively since it is now something that will cause an explosion and a sign of life. The heartbeat-fuse is being used to connote the purported destructive-generative potential of technological innovation. The narrator continues, "Every so often a social or economic upheaval would change the way people thought, connected, and did business" and the words SOCIAL and ECONOMIC pop out the fuse. The screen then goes to gray and the words CHANGE, THOUGHT, CONNECTED, and then BUSINESS

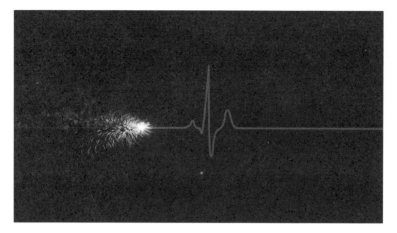

Figure 5.3 *Disruption as burning fuse/heartbeat.*

appear at the center of the frame. The burning fuse then returns and, true to technological determinism, the narrator then states, "Then along came the Internet and, BANG!, everything changed." Just before the narrator says, "BANG!," WWW appears under the fuse and explodes with the "BANG!" The screen then fades to white and the words EVERYTHING and CHANGED appear to underscore the words of the narrator. Both verbally and visually, the Internet, or more specifically the WWW, is being attributed with this disruptive power to abruptly change how people think, communicate, and act.

In keeping with technological determinism, the Internet is conceived of as a technical "thing" that enjoys a kind of autonomous existence. It arrives causa sui without a history and profoundly changes the social world with which it collides. Rather than understanding the Internet as an assembly of technologies, practices, and knowledges that evolved within a particular social milieu, it is, instead, treated like a quasi-object that collides with the existing social order, producing a new one. The effects of the change brought on by the new innovation are treated as though they are the inevitable outcome of, and intrinsic to, the technology itself. This depends upon a mechanistic understanding of causality in which, like Newton's Third Law, "For every action, there is an equal and opposite reaction," the arrival of a new technology acts upon and directly produces societal (or economic) effects.

Like the Siemens article, the video also draws upon many of the same resources of technocratic discourse that Graham identifies in order to make its claims about how Australian businesses should respond to "digital disruption." For example, the narration makes use of process metaphors, which Graham (2001: 768) explains are a kind of grammatical metaphor in which the represented process does not fit neatly into one of Halliday's six types

Figure 5.4 *Explosion as a process metaphor.*

of experiential (material, mental, verbal, relational, behavioral, and existential) processes. In such instances, rather than functioning as one of six realms of human experience, "the process retains its grammatical standing *as* a process, but functions very differently" so as to "imply action *throughout the whole of human experience all at once*" (Graham 2001: 768). Thus when the narrator makes the statement, "Digital disruption explodes the status quo, rewrites economics, scrambles supply chains, and blurs category boundaries," we can see how "explodes," "rewrites," "scrambles," and "blurs" all seemingly function as material processes, but in actuality the processes constitute a relationship between "highly condensed, extremely abstract nominal groups that compress myriad, complex and massive processes into static things" (Graham 2001: 768). By using a process metaphor, digital disruption becomes visible as a real and fact-like thing that is able to cause real, material effects in the world.

I should like to propose that process metaphors can also be realized visually as well as verbally. One thing immediately apparent in the Deloitte Australia promotional video is that there are few actual concrete things being represented. This is because the majority of the nominal groups being discussed in the video are, in fact, also condensed and abstract, and difficult to reduce to a simple visual image. Accordingly, there is considerable reliance upon using lexical and abstract iconic objects in the video as I have already suggested above. The use of the burning fuse/heart trace to represent digital disruption, the clocks to represent business sectors, explosions to represent disruptions, or a growing plant to represent how "Digital lets you target new sectors and customers in a granular way" all exemplify the way the video works to represent the highly condensed and extremely abstract concepts and processes as real, measurable, material objects that can be represented on a "digital disruption index." As Graham (2001: 769) observes, this is "the

Figure 5.5 *Market growth as a process metaphor.*

aesthetic ruse of a process metaphor: when deployed, representations of *irrealis* states look like (are presented as *if* they were) *material actions* in the here and now, as actually existing matters of fact."

The two texts "Coming Soon" and "Digitial Disruption" exemplify the way technocratic discourse presents demands for action in the guise of statements of fact. Both rationalize claims to truth and thereby compel audiences to comply through the use of tense structures, process metaphors, and nominals. The imagined future states that the two texts present are offered as utopian spaces in which market value is maximized. By drawing upon a discourse of technological determinism, they are able to substantiate their claims of future utopian states by pointing to technologies already available in the present that will presumably spur on the users and herald the arrival of these utopias. Technologies such as CMC are represented as discrete things and the cause of change. The use of the letters WWW to represent the difficult to represent Internet typifies the way otherwise complex and abstract processes and activities can be semiotically distilled down into pseudo-objects. Furthermore, these technologies are always represented as if "Bang!" they just happened out of the blue. Even in the case of the workstations, there is little attention to the selective realization of goals, motivations, and organizations that are embedded in those technologies.

Conclusion

Technological determinism treats technologies as finite, isolatable things that autonomously and spontaneously bring about changes to the societies in which they are introduced. The discourse of technological determinism affords laying the blame with, or praising, a technology for the changes it supposedly brings about. In this way, technological determinism allows people to describe technologies as a force for either positive or negative change. It privileges technology as the principal driver of social change and all explanatory power rests with the technology.

The video "Look Up" captures perfectly this notion that the arrival of a new technology will suddenly and profoundly change society. No one, it seems, can resist the small screen and the constant remote sociability that it seems to promise. The only solution is this: break free of the technology, leave it at home, and go out into the world in a more natural, less technological, way. Likewise, the humor of the Coca-Cola social media guard is that it takes a technology to defeat a technology.

If technological determinism overstates the role of technology in society, the question that needs to be posed is this: How do we acknowledge the significance of technology without overprivileging it? Again, the answer lies in articulating an alternative discourse that eludes the hylomorphic trap. The concept of the associated milieu at once affords a way of conceptualizing both the agency of the technology and its milieu or socio-technical environment since the technology must be concretized in a particular environment, and while this brings changes to the environment, the environment also conditions the becoming of the technology.

6

Discourses of technological fetishism: (Over)valuing technologies

Technological fetishism is obviously closely related to technological determinism. Both share in the assumptions that technologies are discrete, finite things that create real and profound effects in the world, and do so spontaneously and independently. However, technological fetishism should be reserved for a specific variant of technological determinist discourse. While it is quite possible to attribute causative powers to a technology independently of the broader apparatus that is organized around it, technological fetishism depends upon a process of investing social value in the technology in a way that is not intrinsic to technological determinism. What I want to argue is that technological fetishism arises from overvaluing technological objects beyond simply attributing independent and determining agency to them.

Technological fetishism depends upon a sublimated overinvestment in the technological object. Technological fetishes are constituted through a threefold process. First, there is the generation of written texts, images, and discussion about the object itself such that the object is put into discourse. Second, the object is situated within a series of relations between other objects and people out of which emerges the structural effects of a network. And third, there is the investment of values and capacities in the object so as to give it profound social significance or what de Sassaure would have called valeur. Together, these three processes invest the values, obligations and capacities, and knowledges in the technical object in a manner that is overdetermined (Dant 1999).

In this chapter, I propose to demonstrate how discourses of technological fetishism are realized by the overrepresentation of social value in technical objects. The chapter starts by considering how the experience of the sublime was attributed to technology and the intensification of value that it realized. The technological sublime is closely related to technological fetishism, but like progress, it does not hold the same currency since technologies themselves have come to be experienced as more mundane and everyday than in the nineteenth century. As a result, the sublime itself has come to be secularized as demonstrated in an account of the opening to the US Army video game, *Future Force Company Commander*. The chapter then uses the example of Explosive Ordnance Disposal robots to explore how the realization of the discourse of technological fetishism extends beyond technological determinism through the overvaluing of technological objects.

The technological sublime

The technological sublime refers to the way in which in the nineteenth century, industrial machines came to be understood as both objects of wonder that carry unsettling power and signs of progress. The concept of the sublime has a long history in philosophy and the notion of the sublime as inspiring an experience of awe and feelings of reverence can be traced back to antiquity. While initially applied to rhetoric, in the eighteenth century, it began to be applied to natural objects such as mountains. The sublime, with its capacity to inspire awe constitutes an excess and, for eighteenth-century philosophers like Burke, represents the inverse of Enlightenment reason: "a permissible eruption of feeling that briefly overwhelmed reason only to be recontained by it" (Nye 1994: 5). The technological sublime describes precisely how the industrial age machine was able to simultaneously be experienced as awe-inspiring and as the materialization of rational progress.

The sublime itself describes a feeling of having encountered a sign of the perfection of creation, of the omnipotence of God, which is beyond the capacity of mere mortals to fully understand or contemplate. What brings us to the sublime is experienced as both wondrous and overwhelming. The sublime, therefore, describes an encounter with the divine mediated through a material form here on earth. As Nye (1994: 3) argues, though, where we encounter the sublime is open to change: "The history of the sublime from antiquity shows, if nothing else, that, although it refers to an immutable capacity of human psychology for astonishment, both the objects that arouse this feeling and their interpretations are socially constructed." The technological sublime, therefore, represents an (re)investment of this sense of glimpsing into the perfection of creation so that the machine has the potential to bring on this experience of bewilderment and awe. As Nye (1994: xiii) proposes, in response to "a physical world that is increasingly desacralized, the sublime represents a way to reinvest the landscape and the works of men with transcendent significance."

The technological sublime as a reinvestment of fervent religious experience in technology is typified in the opening sequence of *Future Force Company Commander* (*F2C2*) (Zombie Studios Inc 2005). F2C2 is a real-time strategy game developed by Zombie Studios on behalf of SAIC and the US Army. The game was intended to showcase the canceled Future Combat Systems (FCS) Army modernization program in which SAIC and Boeing acted as lead contractors. Like the game *America's Army* (Zombie Studios Inc 2002), *F2C2* was also intended to act as a recruitment tool as well as a promotional tool for the costly FCS procurement program. Gaming critics frequently noted that, as a game, it was relatively easy to win and that it seemed to function more as a promotional tool than an actual game.

The game included a two-minute computer-animated (CGI) opening sequence in which an enemy airbase is bombed and then secured, and a potential enemy counter-attack is thwarted. The sequence demonstrates how an FCS assault would unfold and how the technologies would provide soldiers with "unprecedented situational awareness, and the ability to see first, understand first, act first and finish decisively" (2006). The upshot is that the focus and, therefore, the central protagonists of the narrative are not the US soldiers but their equipment. Presented in a video game format, in which action is typically presented as spectacle, the representation of the FCS systems clearly draws from notions of the technological sublime.

Opening Sequence, *Future Force Company Commander* (*F2C2*)

Shot 1—zoom to Satellite orbiting Earth

Music: pitch low with long attack

Shot 2—zoom downward through clouds toward Earth and low-orbit reconnaissance satellite

Audio transition "whoosh"

Music: pitch rapidly rises in intensity, ends abruptly with short decay

Shot 3—light-flash and cut to polarized POV shot of air base

Audio transition "whoosh"

Shot 4—cut to soldier operating long-range viewing device

Audio transition "whoosh"

Shot 5—light-flash and cut to polarized POV shot of enemy air base—zooms in on enemy base

Audio transition "whoosh"

Shot 6—light-flash and cut to close-up polarized POV of hangar bay door—an enemy fighter jet is parked in front and another is inside. Pan to armored anti-aircraft vehicle and troops.

Audio transition "whoosh"

Shot 7—cut to zoom in on solider again—B2 Stealth bomber flies over head

Audio transition "sound of B2 engines"

Shot 8—dissolve transition to bomber dropping payload of guided bombs

Audio transition "whoosh"

Shot 9—cut to guided bombs gliding downward through clouds

Audio transition "whoosh"

Shot 10—cut to single bomb striking and destroying enemy anti-aircraft vehicle

Shot 11—cut to inside enemy hangar, enemy soldiers run from bay door as bombs fly in, directly striking aircraft outside and inside

Music: pitch rising slowly in intensity with long attacks and short decays

Shot 12—cut to bomb making a direct hit and destroying missile launcher

Music: pitch rising slowly in intensity with long attacks and short decays

Shot 13—cut to aircraft carrier as VTOL aircraft takes off and flies toward combat

Shot 14—cut to soldiers descending by rope from hovering VTOLs

Music: pitch rising slowly in intensity with long attacks and short decays

Shot 15—cut to Hercules-type aircraft armed with 40 and 105 mm cannons

Music: pitch peaks in intensity with long attacks and short decays

Shot 16—cut to sensors being dropped from aircraft

Music: drops suddenly in intensity

Shot 17—sensors strike and self-plant in the ground

Shot 18—cut to soldier watching screen inside the command tank. Images of enemy tanks flash on screen and stop at Soviet-era T-72

> Music: resumes at previous intensity

Shot 19—cut to sensor, column of enemy T-72 tanks passes near sensor

> Music: pitch rising slowly in intensity with long attacks and short decays

Shot 20—cut to squadron of Hercules type-aircraft flying overhead

> Music: pitch rising slowly in intensity with long attacks and short decays

Shot 21—light-flash and cut to level shot of aircraft squadron beginning landing descent

> Music: pitch rising slowly in intensity with long attacks and short decays

Shot 22—aircraft land and light armored tanks drive out of craft

> Music: pitch sustains intensity with long attacks and short decays

Shot 23—cut to soldiers at terminals inside the command tank

> Music: pitch rising slowly in intensity with long attacks and short decays

Shot 24—cut to front of the point tank crossing bridge

> Music: phrasing begins to change with quicker attacks and short decays

Shot 25—cut to underside of bridge as tank passes over

> Music: phrasing begins to change with quicker attacks and short decays

Shot 26—cut to zoom on helicopter-type UAV

> Music: attacks continue to shorten with short decays

Shot 27—light-flash and cut to polarized POV shot of bridge and tanks

> Music: now both attacks and decays are quick

Shot 28—cut to tanks moving across bridge and firing at enemy

> Music: peaks in intensity and phrasing peaks with rapid attacks and decays

Shot 29—cut to enemy soldier with shoulder-fired anti-armor rocket sighting point tank

> Music: abruptly stops with Grand Pause

Shot 30—cut to polarized POV of point tank being targeted (no light-flash)

> Music: returns as short low-pitch pulsing sounds with short attack and decay

Shot 31—cut to enemy soldier firing rocket

> Music: short low-pitch pulsing sounds with short attack and decay

Shot 32—cut to rocket flying toward tank but being destroyed by tank's defensive system

> Music: short low-pitch pulsing sounds with short attack and decay

Shot 33—cut to small UAV launching

> Music: short low-pitch pulsing sounds with short attack and decay

Shot 34—cut to soldier at terminal in command tank—enemy soldier is sighted on the screen

> Music: short low-pitch pulsing sounds with short attack and decay

Shot 35—cut to light armored mortar, which then fires guided round

> Music: short low-pitch pulsing sounds with short attack and decay

Shot 36—cut to mortar round reaching apogee of flight

> Music: short low-pitch pulsing sounds with short attack and decay

Shot 37—cut to mortar round descending directly upon enemy soldier

> Music: Cantata begins with slow rise in pitch and intensity

Shot 38—light-flash and cut to polarized POV shot of enemy base at other end of bridge

> Music: Cantata continues rising in pitch, tempo, and intensity

Shot 39—cut to MULE Unmanned Ground Vehicle (UMG) entering perimeter of base

> Music: Cantata continues rising in pitch, tempo, and intensity

Shot 40—cut to long shot of MULE front-on

Shot 41—cut to low angle zoom in on MULE camera as it pans the battlespace

Shot 42—light-flash and cut to MULE's polarized POV shot (previous POV was therefore also MULE's) of enemy tanks and soldiers

Shot 43—cut to armored artillery battery firing

Music: Cantata builds in intensity

Shot 44—cut to light armored tank firing

Shot 45—cut to rocket batteries firing

Shot 46—cut to munition directly striking and destroying enemy tank behind building

Shot 47—light-flash and cut to polarized POV of enemy tank being destroyed

Music: Cantata reaches crescendo

Shot 48—light-flash and cut to column of enemy tanks as each one is struck and destroyed

Music: Cantata crescendo sustains

Sounds of anvil strikes

Shot 49—cut to zoom out of bridge head as lead tank leaves bridge and small UAV flies across screen

—zoom continues into clouds

—winged UAV flies across screen above clouds

Music: Cantata replaced by low-pitch long decay tones, using synthesizer

Shot 50—cut to back of soldier's head at terminal in command tank

Shot 51—cut to zoom in on central display screen of real-time map in command tank

fade to black

F2C2 represents the military might of the United States as both omniscient and omnipotent. The opening sequence begins with a high-orbit spy satellite and then zooms to a low-orbit reconnaissance satellite. The US forces have total awareness of the battlespace because of their surveillance technologies. Satellites, sighting systems, UAVs, UGVs, and sensors all deliver information to the company commander seated in front of the onboard battlespace visualization system in the command tank. From there the company commander is able to call in salvos from a variety of weapon systems. Perfect strikes are depicted taking place through dense cloud cover, inside hangars, and behind buildings. There is literally nowhere to hide as shown by the guided mortar round being dropped on the enemy soldier who fired on the tank crossing the bridge. Because they have technological superiority, there is no place that is beyond the US Army's visual field and reach such that whatever appears on their screens can and will be systematically blown to pieces.

Interactionally, the video uses POV shots to put the viewer on side with the US Army. The video does include a POV shot from the perspective of the soldier attempting to destroy one of the US light tanks, but this is done to show the viewer that the enemy has targeted and locked on the tank crossing the bridge. This then allows the Army to show how the onboard defensive system would protect the tank and its occupants. Not only the lethality but also the survivability of the US technology is being showcased. Over-the-shoulder shots are also used to show the company commander at work inside the C2V. Such shots allow the viewer to get a sense of how the visualization system might be used in so-called network-centric warfare to organize and act on information collected throughout the battlespace. At the same time, these shots also highlight the technology over the operators. It is the monitors and the kinds of information they display that are the most salient and, therefore, most important in these shots.

Despite being almost entirely CGI, the video affords claims to high modality. This is realized, in part, by the use of photo-realistic images that users of military video games have come to expect. The weather conditions, cloud and snow, also add to the realism of the sequence. But the high naturalistic modality is also realized through the use of a hi-fi soundscape in which sounds are heard clearly and distinctly and there is no competing background noise (see Machin 2010: 217). Sounds such as those of a camera lens, clanking tracks, dropping bombs, and the impact and dirt displacement of falling ground sensors all add to the realist modality claims being realized by the video. Additionally, these diegetic sounds along with the sounds of the B2 engines and the tracks of the rolling tank also provide salience cues for the viewer.

As a composition, the sequence unfolds quickly with 51 distinct shots being included in the two-minute video. Aside from Shots 1 and 2, which use zooms, and Shot 8, which uses a dissolve transition, all of the other edits are cuts with little to no visual transition. Shots 3, 5, 6, 21, 27, 38, 42, 47, and 48 all use light-flash cuts. As editing transitions, they differ from dissolves in that they are of short duration and, like framing, they separate rather than blend the two connected shots. In this way, the light-flash edits provide the viewer with a cue that the ensuing scene should be treated as somehow distinctive. When used in editing, light-flashes have the potential to convey high energy and power, and in all but two edits, the light-flashes mark the switch to a machine vision POV shot. Shot 27 is at the mid-point of the opening sequence and marks when the assault on the enemy air base is about to commence. Shot 48 is near the end of the sequence at the climax of the assault when the

Figure 6.1 *Enemy soldier is sighted.*

Figure 6.2 *Guided mortar round strikes enemy soldier.*

cantata has reached its crescendo. This last light-flash edit not only redounds with the climatic peak in the action and musical score but also potentially ties what is happening on the screen to the visualization systems in the POV shots. Because the light-flashes are used consistently to introduce the enhanced vision produced by the Future Combat System suite of vehicles and sensors, it is reasonable to suppose that the flashes also have symbolic meaning, suggesting immediate illumination, that is, global vision, afforded by the technology. But in addition to piercing through the "fog of war," the flashes also lend an aura of power and spectacle to the way in which information is being captured and relayed through the FCS network.

Sound cues can also be used in video editing to produce textual cohesion. Shots two through to eight use "whooshing" transitions with fast attacks and decays to enhance the experience of action, momentum, and velocity. Attack and decay refer to the way a given sound rises and recedes, respectively (Machin 2010: 111). The whooshes quickly rise and fall in intensity, suggesting a sudden release or expenditure of energy. It also has the potential to realize

Figure 6.3 *Cantata reaches crescendo as enemy tanks are destroyed.*

a sense of cohesion and coordination since each action neatly follows the previous one.

If the technological sublime "is an essentially religious feeling, aroused by the confrontation with impressive objects" (Nye 1994: xiii), then the opening sequence of *Future Company Commander* seeks to arouse such feelings toward the FCS program. On its own, the video differs little from the techno-fetishism that is typical of much of military futurism. It is the musical score and, in particular, the use of the cantata that endows FCS with qualities of the sublime. The opening music works to build tension and create a sense of momentum starting with long attacks and short decays. The attacks then become progressively quicker as the pitch slowly rises in intensity. Just as the enemy soldier fires the rocket at the armored vehicle, there is a grand pause. In musical composition, the grand pause is often used for dramatic effect during a loud section of the piece. Immediately after the pause, the cantata begins coinciding with the targeting of the enemy soldier. Ultimately, the choice of a cantata, with its associations with Christian religiosity, redounds with the god-like power that technology is depicted as granting the US army.

Conceptualizing fetishism

Fetishism as a concept enters into social theory in the proto-anthropological accounts of "savage" peoples and their supposed practices of worshiping objects as a type of pre-religion. The origins of the term fetishism are exceedingly well established and, indeed, Masuzawa (2000: 243–44) notes:

> A typical Victorian account of fetishism would rehearse the etymology of the word, in the course of which we are transported back to the scene of

the first encounter between Portuguese sailors and savages of the Gold Coast. At this point we would be led to examine the Portuguese word *feitiço*, meaning "charm," "amulet," or "talisman," which in turn might lead us back through medieval Christian history to a Latin term *factitus*, meaning, variously, "manufactured," "artificial," "enchanted," or "magically artful." Then the narrative would likely go forward to 1760, when the French Enlightenment thinker and acquaintance of Voltaire, President Charles de Brosses, coined the term "fetishism" in the now celebrated monograph, *Du culte des dieux fétiches*.

From this proto-anthropological conception, fetishism has, of course, been incorporated into psychoanalytic and Marxist theory to characterize the relationship between objects and those that desire them. Both Freud and Marx, in inheriting this conception of fetishism, attribute to misunderstanding of the nature of the significance of the object and the misattribution of qualities properly associated with human subjects to objects instead. For Marx, the fetishized commodity has magical properties by virtue of its ability to make "the social character of men's labour" appear "as an objective character stamped upon the product of that labour" (Marx 2005b: para 4). While for Freud (1975: 154), the fetish object allows the fetishist to accommodate the "fright of castration" by imagining the object to be a substitute for the penis that women would otherwise lack. As such, the significance of the fetish object is arrived at by being endowed with a kind of sign value. Marx (2005b: para 7), for example, describes the fetish object as being a "social hieroglyphic" that "is just as much a social product as language." In each case, the fetish functions on the basis of a value being assigned to the object that is derived from its ability to signify something else of greater value (social labor or the phallus). This also means, though, that the value of the fetish is misattributed and the label of fetishist is applied to those who would misunderstand the true nature of the object. Within rational-scientific discourse, then, fetishism is always a characteristic of the thinking of others. In being used to pass judgment on a perceived overvaluation of an object that is indicative of a false consciousness of the real conditions of the world, fetishism would seem to be of limited usefulness in thinking through the actual processes of valuation that endow the object with its special significance.

In his critique of the Marxian treatment of the commodity, Baudrillard (1981: 90) has noted: "The term 'fetishism' almost has taken on a life of its own. Instead of functioning as a metalanguage for the magical thinking of others, it turns against those who use it and surreptitiously exposes their own magical thinking." For Baurdrillard (1981: 88), evoking fetishism is a kind of fetish itself: "It is the conceptual fetish of vulgar social thought." Turning to what Baudrillard refers to as the "great fetish metaphor" is dangerous because it forestalls analysis by substituting it for a variant of magical thinking about the thinking of others. As such, the metaphor of fetishism depends upon and at the same time shores up a rationalistic metaphysics. By finding irrationality in the thought and error in the judgment of others, one can readily believe that one's own thought is both rational and sound. At the same time, however, I would argue that this does not entirely preclude using the concept of fetishism within critical thinking about the magical thinking of others.

Despite his forewarning of the fetishism of fetishism, Baudrillard opens the door to a more productive or "positive" conception of fetishism. In his etymology of the term fetish, Baudrillard also points to the earlier meaning of the term that precedes the "semantic distortion" that has transformed our contemporary comprehension of the significance of the fetish to that of a misattribution or misunderstanding: "Today it refers to a force, a supernatural property of the object and hence to a similar magical potential in the subject. . . . But originally it signified exactly the opposite: a fabrication, an artifact, a labor of appearances and signs" (Baudrillard 1981: 91). From this Baudrillard (1981: 93) concludes that "if fetishism exists it is thus not a fetishism of the signified . . . it is a fetishism of the signifier." The sign value of the fetish as a social hieroglyphic comes not from obscured or obfuscated social labor but rather from the very social labor of fetishizing itself. It is this observation that suggests another more "positive" or productive understanding of the process of fetishism. If fetishism is understood as "a cultural sign of labor" (Baudrillard 1981: 91) rather than simply as a mistaken belief in the order of things, then the concept of fetish might still serve the purposes of critical theory by pointing to the role of the fetish object as a "mediator of social value" (Dant 1999: 41). It is this role as sign or mediator of social value that begins to explain the fascination the fetish exercises over the fetishist.

Accordingly, I am arguing that the proto-ethnographic conception of fetishism does not preclude using the concept of fetishism to critically think about the magical thinking of others. Hornborg (2011: 22), for example, draws from Latour's call for a symmetrical anthropology to propose an "anthropology that does not merely represent an urban, 'modern' perspective on the 'pre-moderns' in the margins, but that is equally capable of subjecting modern life itself to cultural analysis." For Hornborg, fetishism is not simply a false understanding of subject-object relations but rather constitutes a strategy for knowing the world. What Hornborg is pointing to is a more "positive" or productive understanding of the process of fetishism. Likewise, for Dant (1999: 121), the fetish object is understood as a site of mediated social value such that its perceived special qualities are emergent through a process whereby "objects are noticed, are given attention, are drawn into relevance and constituted as meaningful through social interaction." In this way, the fetish object is not simply a negation of a repressed or misunderstood reality, nor is it simply "reflecting back the ideas and beliefs of its worshippers, it is transforming them or, in the language of Actor-Network Theory, 'translating' them" (Dant 1999: 44).

Thus, the fetish is actually more than a sign—like any medium, the materiality of the fetish does not simply "carry" the sign or significance of its value but rather embeds itself in its "message," (re)articulating it so as to privilege a particular trajectory out of many possibilities (see Akrich and Latour 1992: 259). In view of this, the fetish has intersubjective consequences for the fetishists. My intention, therefore, is not to simply adopt a celebratory stance toward fetishism in general but rather to reappropriate the term so as to critically engage with the magical thinking that gives the technical object its fetish status. We may at some level all be fetishists, but what we choose to fetishize is telling.

For an object to acquire the status of fetish, it must be transformed into a sign of cultural value and this transformation is accomplished through its appropriation into social relations. If we accept Baudrillard's (1981: 63–87) thesis that *both* human need and object use are socially constituted, then the fetish object derives its status not through any intrinsic property but rather through the significance granted to its perceived or attributed capabilities. And just as needs are socially derived, the functionality of the object is not so much an intrinsic property as it is what ties the object to a system of needs and practices that arise in relation to those defined needs. To Baudrillard (1996: 67),

> "functionality" in no way qualifies what is adapted to a goal, merely what is adapted to an order or system: functionality is the ability to become integrated into an overall scheme. An object's functionality is the ability to become a combining element, an adjustable item, within a universal system of signs.

It is the degree to which the technical object can not only offer a "needed" functionality but also exceed it—offer it in excess—that gives it its value and thus allows it to gain the status of fetish. Therefore, in order for the object to be venerated, "it must be valued according to a code of functionality which orders both human subjects and material objects" (Dant 1999: 49). Following from Dant (1999: 56), the concept of fetishism can, accordingly, be extended as an analytic tool to examine how the social value of some objects is overdetermined "through ritualistic practices that celebrate or revere the object, a class of objects, items from a 'known' producer or even the brand name of a range of products."

The enthusiasm for Explosive Ordnance Robots (EOD) during the height of the "IED crisis" in Iraq is an excellent example of technological fetishist discourse as "a labor of appearances and signs" (Baudrillard 1981: 91) rather than simply a misperception of supernatural powers. EOD robots are essentially remote control devices used to either disable or detonate improvised explosive devices at a distance. Though military robots are increasingly being developed with a wider range of sensor devices and "weaponized," these robots are in no way autonomous devices. Nonetheless, much of the discourse addressing these objects often suggests a certain degree of autonomous intelligence and thus, I would argue, a fetishistic valuation of the EOD robot. To illustrate this, I have selected one particularly germane text drawn from a corpus of twenty-five United States Department of Defense press releases and 177 US newspaper articles (primarily feature articles of between 500 and 1000 words) that make reference to the use of robots as part of anti-IED measures being taken in Afghanistan and Iraq. I also reference other texts from the corpus to further substantiate the claims I am making.

Robots as technological fetish

Below is the entire verbal text of "Command Assesses Robot to Help Save Soldiers' Lives" (Sanchez 2005), which is a press release that was published on the website for the United States Department of Defense's (DoD) Business Transformation Agency. Actually composed by a member of the Operational Test Command Public Affairs Office, the release outlines the Operational Test Command's (OTC) assessment of a particular type of robot that was deployed to US troops in both Afghanistan and Iraq. The textual analysis that follows identifies the ways in which the actors are represented within the text and the ways in which their actions are also represented. Of particular interest will be identifying how the lexicogrammar used to represent actors and action contributes to the rhetorical objectives (Lassen 2006: 505; White 2000: 1997) of the text.

Command Assesses Robot to Help Save Soldiers' Lives (Sanchez 2005)

1. It looks like a robot from the movie "Short Circuit"—a miniature Johnny 5 that rolls on four wheels and has a video camera lens for eyes.

2. It peers around doors and windows, carefully adjusts its height to survey the area, and rolls carefully toward suspicious-looking vehicles and objects.

3. There it remains still, focusing on the suspected object, and waits for a signal from the operator to tell it where to go to next.

DISCOURSES OF TECHNOLOGICAL FETISHISM 153

4   The Multi-Function Agile Remote-Controlled Robot, or the MARCBOT, demonstrated its abilities to soldiers from the US Army Operational Test Command on 12 January at the Fort Hood Military Operations on Urbanized Terrain site.

5   With only a little over an hour to have hands-on training to operate the robot, Operational Test Command's test players, who were soldiers from the 1st Battalion, 22nd Infantry of the 4th Infantry Division, were ready to conduct a sweep in one of the urbanized terrain site buildings as though they were in the streets of Baghdad.

6   Capt. Michael Fitzgerald, the command's test officer from the Future Force Test Directorate in West Fort Hood, said that the MARCBOT is one of a platform of robots currently being used in Iraq to do surveillance and reconnaissance of improvised explosive devices.

7   Today's soldiers have had so much exposure to modern technology, such as playing with remote-controlled cars and operating video games, that it doesn't take long for them to adapt to the controls of the MARCBOT, he said.

8   "The Army has already fielded these robots to our soldiers in Iraq because of an urgent need," Fitzgerald says.

9   "The robot has been helpful for our soldiers, but what we don't know right now are its full capabilities and limitations."

10 Fitzgerald said that the data collected from Operational Test Command's assessment could determine whether the MARCBOT design will be redeveloped to add features that could be more helpful for our soldiers.

11 The MARCBOT resembles a large remote-controlled car one finds at the local toy store.

12 It has a retractable arm, all-terrain wheels, and a wireless video camera attached to it.

13 The soldier operates the robot with a remote-controlled operator with a monitor on the controller that allows the soldier to see what the robot sees.

14 "Can the soldier communicate with the robot at great distances?

15 At what range?

16 What are its limitations and what can be improved in the robot to benefit the soldiers even more?

17   These are some of the questions OTC will address during this assessment."

18   Fitzgerald says that arming and protecting every deployed soldier with the best equipment available is the Army's first and constant priority.

19   "Testing or assessing the equipment's capabilities and limitations is Command's priority."

20   "We're the trusted agent for the soldiers in the Army."

21   The MARCBOT is made by Exponent Incorporated and costs around $5,000 each.

22   The operational assessment costs virtually nothing for the Army except for the time invested in practicing and conducting the assessment at the urbanized terrain site, Fitzgerald added.

Representing Actors and (Re)Allocating Agency

While van Leeuwen (2008) restricts his definition of social actors to human ones and, likewise, social action entails the actions of human actors, the transitivity structures used in the press release clearly represent the robotic device in question as an actor in its own right. In particular, clause complexes 1–3, 4, and 9 each represent the robot as an autonomous actor capable of performing actions, cognition, sensing, and communication. This is done in part by the ways in which the robot itself is represented in the lexicogrammar, and also by the options selected to represent the soldier-operators within the system network of the text. It is this distribution of agency between the MARCBOT and the soldier-operators who are actually controlling the robotic devices that makes the MARCBOT seem to be an autonomous device.

The realization of agency attributed to the MARCBOT (robot) as well as the passivation of the soldier-operators can, for example, be seen in sentences 1 through 3, where the MARCBOT is presented to the reader in a clearly animistic fashion and the operator is not present until the third clause complex. By de-agentalizing the action and backgrounding the operator, the MARCBOT seemingly behaves as described under its own will.

Thus, in sentence 1, the robot is compared to the robot character from a movie (I will comment more on that below) and is said to possess the equivalent of eyes as if it is actually able to see. The robot is also attributed with the ability to roll on four wheels and this could be said to be de-agentalized since such an action is not actually a "behavior" of the robot itself but the result of an operator who would control the device and cause it to move by sending the appropriate radio-signals to the device. As such, the lexicogrammar represents the described action as something that happens without human intervention.

In sentence 2, again, the robot is activated as an actor. The robot is represented as being able to "peer around doors," "carefully adjust its height to survey," and roll "carefully toward suspicious-looking vehicles." Clearly, the care taken and the suspicion felt is actually that of the operator that remains unrepresented or backgrounded until sentence 3. While grammatically there is an agent performing the material processes, in terms of congruence to "reality," we can clearly argue that the action has been de-agentalized.

In the third clause complex, the robot operator is actually represented but in a form that realizes passivation. The reader is informed that upon reaching the suspicious object, the robot will remain still, focus upon the object, and wait for a signal. Agency here is still being granted to the robot. Arguably, the collocation "from the operator" functions grammatically as a circumstance and so the operator is, in fact, represented through passivation since the robot will wait "for a signal . . . to tell it where to go." The MARCBOT, therefore, continues to be activated since the actual processes of moving and surveying with the robot remain de-agentalized.

Together, the three clause complexes very much realize a representation of the MARCBOT as a functioning autonomous technology, when, in fact, it differs little from a remote-control toy car, as is suggested in sentence 7 of the release. Furthermore, the use of parataxis to structure the clause complexes further obfuscates causality since the clauses are simply combined using the conjunction "and" without subordination. The process (action) of each clause just seemingly happens independently of the preceding one.

Continuing in sentence 4, the robot is then nominated as MARCBOT and is reported to have "demonstrated its abilities to soldiers." While the MARCBOT is activated as actor, the operator and others putting on the demonstration are again backgrounded. Thus the action being represented is again de-agentalized since no one is actually attributed with having put on the demonstration at this point in the text.

The agency of the robot is also realized in sentences 13 and 14 where it does not actually function grammatically as the agent in the clauses. Instead, the agency of the robot is effected largely through lexical choice and metaphor. For example, in sentence 13, the robot is represented in the nonfinite clause "to see what the robot sees." Here the robot is grammatically part of the goal, "what the robot sees," but the choice of "sees" further reproduces the animism that often characterizes the descriptions of the robots. It could just as easily have been written that the monitor allows the soldier to see what is in front of the robot. Likewise, in sentence 14, though the robot is again the goal, the process opted for is "communicate," which suggests a transactive semiotic process that is interactive rather than an instrumental material one (cf. van Leeuwen 1995: 90–91). Again, the process could, instead, have been "operate the robot at great distances," which would have been nontransactive and instrumental.

The MARCBOT, then, tends to be presented as an active agent, particularly at the start of the text. Though it is, in fact, an inanimate object with no actual artificial intelligence, the MARCBOT is granted a level of agency that goes beyond the actual technology used to create it. This representation of the MARCBOT as an autonomous technology, I argue, does not stem from misunderstanding. No one, including the author of the press release, really believes the MARCBOT to be more than it really is. Instead, as I will discuss below, the animation of the MARCBOT is a product of a socio-technical imaginary that seeks to legitimize warfare by offering risk-transferring solutions to offset public concerns regarding casualties and, ultimately, reducing the political risks of waging war.

In addition to the MARCBOT, the text also allocates a significant role to Captain Michael Fitzgerald. Fitzgerald serves as the spokesperson for the US Army OTC, which supports the US Army's acquisition and development programs by conducting operational testing and analyses. Accordingly, the action that Fitzgerald performs in the text is semiotic rather than material. Fitzgerald is quoted either directly or indirectly from the sixth clause complex onward excepting sentences 11, 12, and 13, where there is no projection or quotation, but nonetheless they are still likely to be paraphrased from the interview of Fitzgerald.

As the spokesperson for OTC, Fitzgerald is the only actor that engages in semiotic rather than material action in the text. His close identification with OTC is signaled by his use of the pronoun "we" in the ninth clause complex. Much of his speech is quoted directly and not behavioralized. As van Leeuwen (1995: 91) notes, this is usually indicative of high-status social actors. Fitzgerald is clearly being consulted as an authoritative source given his nomination including rank or honorification (see van Leeuwen 1996: 53). Furthermore, the style of reporting in this text is very typical of "middle-class"-oriented news reporting where official representatives and experts are referred to specifically while the general public is referred to generically (van Leeuwen 1996: 47). Thus, the specification of Fitzgerald can also be contrasted to the genericization as well as backgrounding of the soldiers since no soldier is ever nominated or quoted. Accordingly, the credibility of the press release is drawn from the representation of Fitzgerald, as a member of Future Force Test Directorate within OTC, and the manner in which his speech reports much of the embedded social action represented in the text.

DISCOURSES OF TECHNOLOGICAL FETISHISM

Embedded within Fitzgerald's speech is a carefully crafted account of why the MARCBOT is already in service and why it is currently undergoing an assessment under OTC. In sentence 6, the reader is informed that the MARCBOT is in use because of "an urgent need." This is a perfect example of exclusion by suppression since there is no trace of the actors (the military typically labels them "insurgents") or their activities that have produced the urgent need (planting "improvised explosive devices"). Only in the sixth clause complex do we find mention of IEDs and they are never actually connected to an actor within the text. Indeed, they may as well grow like mushrooms. Of course, there is no doubt that readers will know how and why IEDs come to be planted in order to ambush American soldiers in Iraq, but this, I would argue, is not actually an example of an exclusion that occurs simply because the reader is presumed to know already and naming the actors responsible for planting the IEDs would be "overcommunicative" (van Leeuwen 1996: 41). Instead, since this is an example of strategic writing and public relations, the exclusion is very much in keeping with the desire to downplay the conflict and difficulty faced by the US military in Iraq. Clearly, the focus of this release is upon the abilities of the MARCBOT, the evaluation by OTC, and the benefit to soldiers.

It is this stress upon the benefits soldiers experience and will continue to experience as the MARCBOT is evaluated that shapes the representation of the soldiers in the text. For the most part, the text tends to present the soldiers as passivated actors. There are four clause complexes within the text where the soldiers are represented as active agents: specifically, 5, 7, 13, and 14. At the same time, there are eight instances where the soldiers are represented through passivation as beneficiaries:

4. "demonstrated its abilities to soldiers"

8. "fielded these robots to our soldiers"

9. "has been helpful for our soldiers"

10. "more helpful for our soldiers"

13. "allows the soldier to see"

16. "to benefit the soldiers even more"

18. "arming and protecting every deployed soldier"

20. "for the soldiers"

The passivation of the soldiers is further achieved through the use of the first person plural possessive pronoun "our" in clause complex 8 through 10.

This, I believe, can be interpreted in two ways. First, there is the obvious connotation that the soldiers are represented as a group over which the Army exercises authority and control and that this relationship requires that the Army also be responsible for the soldiers. Accordingly, we are informed in the reported speech of Fitzgerald in clause complex 18 that "arming and protecting every deployed soldier with the best equipment available is the Army's first and constant priority." From this the reader is reassured that the Army is meeting its obligations. In support of the Army meeting its obligation to the soldiers, in clause complex 19, we can then read the quoted speech of Fitzgerald to learn that "Testing or assessing the equipment's capabilities and limitations is Command's priority." The redundant use of "priority" reinforces both the Army's commitment to its soldiers and what could be called the "chain of commitment" from the Army to OTC. But the second possible connotation is one of an implied mutual understanding between reader and the Fitzgerald/author of the possessive pronoun where both parties share in the claim to the soldiers. As citizens of the United States (the intended audience for the press release), these are *our* soldiers and *we* are also responsible for *their* well-being.

The profession of concern for the soldiers is further exemplified by the intensive attribution, "We're the trusted agent for the soldiers in the Army" presented in clause complex 20. Interestingly, there is no actor actually identified as the "truster" but it is likely that the claim is being made that the OTC holds the trust of the soldiers and should, by extension, therefore equally hold the trust of the American public that they have placed the interests of the American soldiers first.

The final two clause complexes concern fiscal accountability rather than necessarily a need to protect soldiers. In keeping with the inverted pyramid structure of a typical press release, whether or not this last bit of information is communicated to the public is likely of less importance. Nevertheless, in the world of armaments, $(US) 5,000 is a paltry sum and the explanation that the assessment entails very little costs to the Army is hardly apt to stir up any controversy.

Ultimately, what the report does is represent the use of robotic technologies in military contexts as a living life-saving technology—a fetishized technological object with "magical" properties. Underlying this discourse is an expression of a "will-to-power . . . to 'imitate life by mechanical means'" (Noble 1984: 58; see also Armitage 1999), thus making war seem somehow compatible with life. It is this notion that war can be waged while at the same time actually "saving" lives that is being used to re-legitimize warfare, a practice that ultimately devalues life. In what follows, I intend to argue that the representation of EOD robots exemplifies the way in which technological fetishism makes technical objects become visible as signs of social value through their constitution as hyper-functional devices.

Misrepresenting robotic agency?

Accounts of the robots also frequently describe them as objects of affection. It is not unusual for both press releases to make associations between military robots and positively viewed robots of mainstream cinema such as Robby the Robot (Schafer 2003), R2-D2 (Komarow 2005), and Johnny-5 (Sanchez 2005). These sorts of comparisons could be said to be "disarming" since they invite the reader to associate a military device with positive (and nostalgic) fictional robot characters. At the same time, these comparisons are also likely drawn from the practice of soldiers naming their robots. At least one EOD team is said to have named their robot Johnny-5 (Garreau 2007; Klein 2007; Bora 2008; Komarow 2005). Relationships of affection are also affirmed by a newspaper report in which Colonel E. Ward of the Pentagon's Robotics Systems Joint Project Office is quoted as declaring, "Nevertheless, soldiers 'love' the robots" (Boyd 2006). Ward is also referred to in another report as typifying the relationship between soldiers and robots as friendship: "Ward's robots, especially the $7,000 MARCBOT, have become the best friends of explosive ordnance disposal troops" (Fleischauer 2006).

It is likely no accident that the collocation "best friends" is used here since it calls upon the familiar adage about dogs and men. A former Marine sergeant major that ran a robot repair shop is quoted as saying that having an EOD robot is "like having a pet dog" (Garreau 2007). The propensity for authors to take up this metaphor is demonstrated in the *Wired* report "The Baghdad Bomb Squad" with the following quote taken from an account of an EOD mission: "He jogs up to meet the robot, grabs the radio, and scurries back. Rainman rolls behind, antennas wagging like a Labrador's tail" (Shachtman 2005). Likewise, in another report the technical director for robotics at the US Navy's robotics lab in San Diego is said to predict that "U.S. soldiers eventually will work with robots in the same way a hunter works with a bird dog" (Bigelow 2006). Indeed, Robert Moses, senior vice president of operations at iRobot has been recently quoted as saying, "They are like the police dogs. . . . The soldiers end up having a relationship with their robots" (Bora 2008). Clearly, the attributes of a human-dog relationship are being assigned to the human-robot relationship, but by implication, the robot itself is also equally being attributed with those positive qualities associated with the dog as companion-working animal.

Finally, the animism of the EOD robot is further realized by accounts of robots destroyed during disposal missions as embodied mortal beings:

> One arm and a visceral cavity, wide open with its contents scattered about, is the cost of saving Soldiers' lives. But the brain is intact. (Trevino 2007)

> A bomb blew up and a robot lost its arm, but Staff Sgt. Matthew Bingaman came through the blast with all four limbs intact. (Speckman 2006)

Jenny is dead, another casualty in the struggle to stabilize Iraq. A robot used in counter-insurgency missions throughout the restive Al Anbar province, "Jenny," met her fate aiding a team of Marine explosive ordinance disposal technicians trying to disarm a deadly improvised explosive device. (Piazza 2006)

A unit nicknamed "Scooby Doo" earned a check mark on its camera head for each explosive device it succeeded in disarming. When destroyed, its operator had formed a personal attachment to the unit and returned it to the repair shop, cradling it in his arms as if it were a wounded child and asking if it could be fixed. (Barylick 2006)

Accounts such as these, particularly the first two, point to the fact that the EOD robot is being understood as a surrogate for an EOD technician and that the destruction of its artificial body is a reminder that otherwise it would have been an actual flesh-and-blood soldier.

At this juncture, it could be easy to simply dismiss the talk of IED defeating robots as a simple proto-anthropological form of fetishism in which the robot is mistakenly understood as an autonomous being. However, the animism of the EOD robot does not stem from any misunderstanding. No one, including the authors of the press releases and newspaper articles, really believes the robots to be more than they really are. Indeed, both the press releases and the newspaper articles that did describe the robot as autonomously performing its "duties" also invariably describe the robots as being tele-operated. For example, in one report (Howell 2007), a program manager for the Center for Commercialization of Advanced Technology is quoted as explaining that the robots are, indeed, tele-operated but later is paraphrased in the same article as saying, "Soldiers could more easily stay out of harm's way if robots were better at telling their own position, where to go and what to search for." Instead of being mistaken for a sign of misunderstanding or misrepresentation, EOD robot fetishism, following Baudrillard (1981), needs to be contextualized as a sign of social valuation and mediation.

It could be argued that such references to robots as autonomous and intelligent helpmates or as objects of affection draw upon the long history of human fascination with automata. The history of automata certainly suggests that the creating of "autonomous" machines can be dated back at least to the ancient civilization of Egypt where puppet-like devices were installed in temples and said to offer prophesy. However, it would be a mistake to suggest a simple transhistorical interest in automata. Instead, it needs to be noted that between antiquity to modernity, according to Jean Claude Beaune, "the context had changed: the object itself was no longer the most important thing, at least from the technical point of view . . .; rather, it was how the machines worked that mattered, their function . . ." (Beaune 1989: 431). Likewise, Noble (1984: 57–58) has proposed that such "delight in automaticity" can also be understood as a "will-to-power" that is, in fact, exercised as the extension rather than abandonment of human control. As both Beaune and Noble suppose, the importance of the automaton has become its labor or rather the possibility of transferring human labor to the device rather than pure wonderment at the artifice. It is precisely the possibility of human labor transference that obviously motivates the adoption of EOD robots. However, it is not just human labor that is being transferred to the robot but also, given the nature of the labor, risk. And it is this displacement that leads us to our alternate interpretation of the process of fetishization.

The fetish value of robots

In order to describe how the EOD robot is made to be visible as a fetish object, I shall review three common and related themes that consistently recur when reporting on military robots: robots (1) saving lives, (2) keeping soldiers out of harm's way, and (3) accumulating risk. These themes, not surprisingly, can be found in both press releases and journalism during the period covering the IED crisis from 2004 to 2007. As Piers Robinson (2004: 105) has noted, activities such as the production of press releases during wartime are a central part of a larger strategic activity of "perception management" on the part of the state to "encourage the development of common media frames over time" and, at the same time, tactically "serve to minimize coverage of damaging or hostile stories." In effect, the press releases work to capture existing popular sentiments regarding robots and ally them to the purposes of war legitimation. It is the repeated extolling of the power of the robot to save lives and keep soldiers out of harm's way in a seemingly autonomous fashion that leads to the veneration of the EOD robot as a fetish object.

Saving lives

>Their ground-breaking work has helped to **save lives** and improve the effectiveness of our military effort (Sestak 2008).

>The Army argued that it needed the new robots quickly, to **save the lives** of US soldiers fighting in Iraq and Afghanistan (Bray 2007).

>"This war has proven that small, fast, easy-to-operate robots **save lives**," Quinn said (Krasner 2007).

>Robotic Contract Awarded to **Save Lives** (US Army 2007b).

> The robot's 2-metre arm used to detonate explosives is critical in **saving soldiers' lives** . . . "I love the idea of **saving lives** with robots," Leonard said (Chun 2006).
>
> "The robot can climb, look inside vehicles or scan ditches. **It saves lives.** . . . What did it do? It gave us 300 more metres. **And saved more lives.** . . . Then they can come home, undamaged, and **live enjoyable lives**" (Fleischauer 2006).

Out of harm's way

> "These war robots, or PackBots, **save soldiers from harm's way**" (Bora 2008).
>
> The robots have since become critical companion tools for U.S. forces, allowing them to complete dangerous missions while **keeping the warfighter out of harm's way** (2007).
>
> Odierno's order specifies that engineers can only detonate relatively simple IEDs, using devices such as robots and the robotic arm of Mine Resistant Ambush Protected vehicles—machines that **keep troops out of harm's way** (Morrison and Eisler 2007).
>
> Redstone's robotics chief strives to put technology, **not soldiers in harm's way** (Fleischauer 2006).
>
> "If we can **put technology in harm's way** and still meet the mission, that's important" (Fleischauer 2006).
>
> "We're putting **fewer soldiers in harm's way**," says Francis, director of the Joint Unmanned Combat Air Systems (J-UCAS) (Silverberg).

As such, the EOD robot is extolled for its ability to meet an urgent need—namely, to save US soldiers by going into harm's way and supposedly reduce the IED threat. Not surprisingly, one of the key rhetorical objectives (Lassen 2006: 505; White 2000) in the press releases that praise the robots is to demonstrate that the military is concerned for the safety of its troops. Accordingly, authors frequently refer to how the military is seeking to provide soldiers with robotic devices in order to reduce the threat of IEDs:

> "Help the Department of Defense and the U.S. Army make robots a bigger reality in warfighting so we can better protect our Soldiers when we send them into harm's way," Bochenek concluded (US Army 2008).
>
> While major new FCS systems may not be fielded until 2012 with the new FCS Brigade Combat Teams, General Casey pointed out that a number of

new technologies "spun out" of the research are already helping soldiers today in Iraq and Afghanistan (US Army 2007a).

The Future Combat Systems also is fielding robots that can save lives (Garamone 2007).

Before the Sept. 11, 2001, terror attacks, whenever the Army needed to fly or tread into dangerous territory, the lives of helicopter pilots or combat soldiers had to be risked. Today, the Army is just as likely to send a robot or flying drone to do the job, and engineers in Huntsville—on Redstone Arsenal, in Cummings Research Park and at local universities—are largely responsible for developing the life-saving technology (Spires 2006).

Thus, it is the stress upon how the soldiers benefit by being provided with EOD robots that shapes the soldiers' representation in the robot reports. By representing the officials as authoritative individuals and representing the soldiers as a collective and passive beneficiary of the robotic technologies being fielded to them, the military is in turn represented as an active institutional actor—one that is looking out for the best interest of its charges.

It is also worth noting that this passivation of the soldiers is periodically further achieved through the use of the first person plural possessive pronoun "our" such as in the quoted speech by Bochenek: "So we can better protect our Soldiers when we send them" or that by Fitzgerald: "The Army has already fielded these robots to our soldiers in Iraq because of an urgent need." The use of such pronouns can be interpreted in two ways. First, there is the obvious connotation that the soldiers are represented as a group over which the Army exercises authority and control and that this relationship requires that the Army also be responsible for the soldiers. But the second possible connotation is one of an implied mutual understanding between reader and the speaker of the possessive pronoun where both parties share in the claim to the soldiers. As citizens of the United States (the intended final audience for the press release), these are *our* soldiers and *we* are also responsible for *their* well-being. It is this second connotation that is most relevant to the concept of risk-transfer warfare to be addressed below.

Accumulating risk

The selling feature of the robotic devices to both the military and the public is the apparent ability to save their soldiers' lives by acting as proxy when confronting a potential explosive device. This is the real labor that the robot has taken on from the human. The value of the robot, therefore, lies in part in its ability to take on and perform transferred labor that is understood as too high in risk to be extracted from valued or "precious" humans. Thus, the valued semiotic labor is not actually the defusing or "controlled detonation" of the bomb—after all, humans can do this too and there is surprisingly scarce attention paid to the actual success rates of the robots—but rather the labor of accumulating the risk that the soldiers would otherwise bear. What the three themes point to then is the *representation* of a process of risk transference from soldier to machine:

> "Before, EOD had to visually inspect an IED, now they send a robot to do that," Mufuka said. "If a robot gets blown up, you can replace it. If a robot arm gets blown off, you can replace it, but you cannot replace a human" (Trevino 2007).
>
> The systems worked together to increase efficiency and mitigate risks to the soldier (Baker, III 2007).
>
> A soldier with a LandShark can remotely detonate a bomb or probe a roadway for booby traps, without risking his life (Bray 2007).
>
> "We want unmanned systems to go where we don't want to risk our precious soldiers," said Thomas Killion (Boyd 2006).
>
> "What's the answer to a suicide bomber? It's doing things remotely," Greiner said. "You're making a decision to put a robot at risk" (Chun 2006).
>
> Now, rather than put a soldier at risk, you can put a robot at risk (Jones 2006).

At first glance, the idea of substituting machines for humans to perform dangerous tasks might seem a noble use of technology. However, this transfer of risk from human to machine needs to be contextualized within relatively recent changes in the way in which Western nations have increasingly come to wage war.

The EOD fetish and its associated milieu

Martin Shaw (2002, 2005) has introduced the concept of risk-transfer warfare to describe the way in which media-wary Western governments have sought to conduct warfare while minimizing the domestic political risks of waging war. Risk-transfer warfare is, for Shaw, a form of warfare in which the risks of warfare are literally transferred over to the other side and which is, accordingly, marked by gross asymmetries of casualty rates between Western forces and local combatants and civilian populations.

Risk-transfer warfare is accomplished through the adoption of technologies that assure the asymmetric distribution of lethality and survivability coupled with careful media management. With the goal of making war acceptable to domestic audiences, Western powers are transferring risk away from their own military personnel and thus avoiding the problem of body bags appearing in the mainstream Western media. Thus, Shaw (2002: 349) concludes, there is a very close relationship between this new way of fighting wars and the new PR-savvy ways of managing news media and public opinion. It is for this reason that Shaw (2002: 349) proposes that the more appropriate term should actually be risk-transfer militarism rather than risk-transfer warfare. Risk-transfer militarism while rhetorically concerned with casualty avoidance, nonetheless, does not treat all bodies equally.

> The care taken for civilians is not only *less* than the care taken for American soldiers, it is *undermined* by a policy adopted to keep the latter safe. Risk to civilians is reduced not as far as practically possible, but as far as judged necessary to avoid adverse global media coverage. *Civilians' risks are proportional not to the risks to soldiers ... but to the political risks of adverse media coverage.* (Shaw 2002: 355)

Thus, risk-transfer militarism is guided by calculations of the risk of injury or loss of life to civilians and Western military personnel only in so far as there is a political risk that has already been calculated.

The valuation of the EOD robot being presented here needs to be contextualized in terms of risk-transfer militarism. It represents the sort of attention to media management and public communication that is endemic to risk-transfer militarism. The underlying purpose of the press releases as strategic writing is to reassure domestic audiences that the military and ultimately the US administration is working in the best interests of the soldiers. The use of robotics is represented as yet another technological solution to reducing American military casualties and therefore is, in fact, part of a broader program designed to secure the continued consent of the American public for the deployment of US troops in Iraq. Thus, by representing robots as taking the place of soldiers in harm's way, the rhetorical objective of assuaging concerns that US troops are being used as IED fodder is being accomplished. The claim can be made that while the US administration is, indeed, deploying its forces in Afghanistan and Iraq, it is not necessarily putting those troops in harm's way.

Furthermore, the veneration of the robot as a solution to IEDs itself represents another key aspect of risk-transfer militarism. There is a second suppressed transfer of risk that is happening here. The use of robots may keep US soldiers out of harm's way, but they certainly do not offer a similar proxy role to Afghani and Iraqi civilians. The robots themselves do not seem to be reducing the numbers of IEDs being planted by forces opposing the foreign military forces present in Iraq. According to a *Boston Globe* article, "IEDs remain the biggest killer of US troops, and have become a key weapon used against Iraqi security forces and civilians" (Bender 2006). Thus, the transfer of risk is not simply to the robot but risk is also further transferred over to the less well protected local military and civilian population as American soldiers become more protected. So while a robot may be able to reduce the immediate risk faced by soldiers upon encountering an IED, they have done little to protect unfortunate civilians from the actual detonation of IEDs. In protecting US soldiers against IEDs, the US military has, in fact, placed the local populace at greater risk since the opposing forces will increasingly opt to use IEDs against "softer targets" as they become less effective against the US troops.

In the case of the robots, they become objects of fascination in the context of being attributed with a "needed" functionality—namely, the perceived ability to defeat the IED threat in Iraq and Afghanistan and therefore reduce US casualties and ultimately the political risks for the War on Terrorism. Understanding how the robots function as a fetish lies not in simply pointing to a misunderstanding that leads to representing them as an autonomous technology but rather the way in which the robots become signifiers of social value through their material and semiotic labor. Just as the television set has, for Baudrillard, a sign value that is in excess of its functional capacities, the robot is also endowed with such sign value through its animistic representation, ability to save lives, keep (US) soldiers out of harm's way, accumulate risk, and, unacknowledgedly, transfer risk. In this way the robot both meets and exceeds the code of functionality in which it is embedded. As such, the fetish value of the robot is overdetermined through a kind of worshipful attitude toward the object "in which reverence is displayed through desire for and an enthusiastic use of the object's properties" (Dant 1999: 58).

Like the IED, there are varied attributes and different morphogenetic histories for each subspecies of EOD robot. The EOD robot is concretized in relation to the milieu of risk-transfer warfare and insurgent counter-strategies of improvised explosives. With its individuation as a technical object of risk-transfer warfare, it comes to be associated with the milieu in which it is operated. Each variant of the EOD robot is never finalized and, in fact, always represents the possibility of further mutations or becomings in association with that milieu. However, by representing EOD robots through the discourse of technological fetishism, the relationality between the robot as concretized object and the associated milieu comes to be disassociated. The EOD robot, rather than being understood as an individuated object in the process of becoming, is, instead, constituted as an individual tool that not only singularly brings about changes in keeping with technological determinism but is also able to act magically in an autonomous fashion.

Conclusion

I have argued in this chapter that technological fetishism is an intensification of technological determinism but also constitutes a distinct discourse about technology. Technological determinism overstates the capacity of technologies to influence the societies that create and use them, but it does not need to elevate the technology to the status of fetish in order to do this. For this reason, I want to reserve technological fetishism for those forms of determinism that overvalue the technology.

Technological fetishism does not just treat the technology in question as autonomous and technology in general as the principal source of societal change. Indeed, technological fetishism does not depend upon espousing technological determinism across the board. Conceivably, a technological fetishist could adopt a determinist view of the fetishized technology alone—thus making it all the more significant. Clearly, it is the labor of investing social value in the technical object that distinguishes technological fetishism from technological determinism.

Technological fetishism disavows the cultural processes of individuation and concretization, which believe the technical object to be purely a matter of technical development. Countering this conception with a more well-rounded and robust account of the concretization of the fetishized technical object would go a long way to challenging the way techno-fetishism de-contextualizes the technology in question. By accounting also for those processes such as cultural, economic, social, and material, which also become concretized into the associated milieu in any technical development, the invisible labor of fetishization can be rendered visible.

7

Discourses of technological (dis)satisfaction: Consuming technologies

In the past, progress served largely as a grand metanarrative (Lyotard et al. 1984) used to legitimate the status quo in the name of an immanent salvation brought to earth through technology. Of course, as Noble (1984: 23) posits, not all were convinced of the promise of progress and did express feelings of ambivalence toward the techno-utopias it promised:

> This is not to say that everyone now actually believed in progress. People still continued to have their doubts about this peculiar and alien notion, and subtly expressed it whenever they talked about such progress: "That's progress, I suppose (isn't it?)" "Well, I guess that's progress (isn't it?)" "You can't stand in the way of progress, anyhow (can you?)" The elliptical questions could still be heard, addressed to some absent authority who presumably knows about such things. Yet, even with their barely audible doubts, and even when progress looked pretty grim in the present tense, people were encouraged by social pressure: to be respectable, to try to be taken seriously, to look progressive.

What Noble is pointing to is a certain degree of ambivalence that people often feel toward technology. On the one hand, there is still the sense that technology is a means to the good life, but on the other, there is increasingly a sense that technology or, better yet, those that own it, might not always work in our own interests. This is particularly the case with the growing debate around technological unemployment and the rise of new forms of automation coupled with Artificial Intelligence.

While technological progress may have once enjoyed the status of "god term," talk of progress in relation to technology has been supplanted largely by notions of convenience and utility. Today, appeals to progress are highly unlikely to solicit the same sense of Durkheimian altruistic sacrifice that they might have done once. For example, Slack and Wise (2005: 19) observe that "it would seem quaint or old-fashioned, to defend one's purchase of a new car as 'progress'." Instead, they propose that "we are now inclined to purchase technologies, not for a sense of the progress of civilization or for the appreciation of grandeur, but for their contemporary manifestation" (Slack and Wise 2005: 19). Slack and Wise refer to this as the mini-sublime. Not the overwhelming and awe-inspiring sense of something greater than oneself but, instead, an aesthetic experience that is tied to self-expression through personal consumption—the sense of "the 'cool' and the 'neat'" (Slack and Wise 2005: 19). It is the experience of finding that perfect little thing but, of course, it is only perfect for now and its aura will soon diminish. Rather than technology being connected to grandiose notions of progress, it is connected to the personal and the promise of personal satisfactions.

Accordingly, this chapter addresses discourses of technological consumption and the satisfactions they promise. The chapter begins by problematizing the notion that technologies afford us labor savings and convenience and argues that the typical way that we experience technology continues to render it black-boxed. The Project Tomorrow after-show at Spaceship Earth in Disney World's Epcot is then offered as an example of how technology has been rearticulated from a civilizing force to a source of personal satisfaction. This notion of personal satisfaction and convenience is further discussed through the example of a promotional campaign for domestic vacuum robots. Finally, through a discussion of the representation of robots as workers, the ambivalence toward technology and how it is negotiated is discussed.

This technology is for you, personally

The shift from progress to convenience has meant a greater emphasis on the personal and private experience of technology. Instead of being encouraged to imagine that a particular purchase would demonstrate that one was a part of progress, consumers are, instead, encouraged to buy goods in terms of what they will offer them personally. Thus, the concept of the sovereign-consumer, endemic to consumer culture, finds expression in what Kress and Adami (2010) call "the ideology of choice." In making choices according to personal interests, the individual consumer can hope to realize a more convenient existence. This is precisely what Project Tomorrow promises is just a day away.

Project Tomorrow is the interactive after-show that Disney World Epcot Park attendees can utilize after riding the Spaceship Earth attraction. While Spaceship Earth depicts the history of media communication as being marked by key moments in technical innovation leading to the rise of Western Civilization and culminates in global digital communication networks bringing us to "the brink of a new Renaissance," the after-show, instead, focuses upon how technological innovation has immediate benefits for the individual everyday lives of the park guests. Riders are transitioned to Project Tomorrow by being asked to complete an on-screen quiz as the vehicle they are riding in descends back to ground level. As part of the pre-recorded voice-over narration, Dame Judy Dench announces:

> For the first time in history, all of us can have a say about the kind of world we want to live in. The choices we have made for the past 30,000 years have been inventing the future one day at a time. And now, it's your turn.
>
> Let's have some fun creating the future, shall we? On your computer screen, answer a few questions for us. Then, we'll show you a new world, custom made just for you. Ready?

The quiz comprises seven questions and takes you to one of four lines of questioning depending upon which answer you select for the first question. The following is the line of questions that proceeds from answering "Home" to the first question.

What are you most interested in?

Home ◉
Work
Health
Leisure

What is more important to you?

The latest technology
Conservation and nature ◉

Where would you like to live in the future?

The city
The country ◉

Where would you rather be?

The mountains ◉
The beach

What would you like your house to be made of?

Recycled materials ◉
All-natural materials

What would you like to grow in your garden?

Food ◉
Flowers

How do you want to travel in the future?

Carpool
Bike ◉

All the questions are directed at "you" and seek to determine your personal preferences in terms of the future. After sufficient time has passed for the quiz to be completed, the narration resumes: "Well done! Now along with your answers let's add in some amazing new technology that we happen to know about," which is accompanied by a series of loaded words associated with technology imploding from the margins to the center of the screen, including "Sensors," "Universal" "Networking," "Wireless Communication," "Smart Card," "Nanotechnology," "Biotechnology," "Telematics," and "Virtual Reality." But there are also words included that pertain more specifically to the personalization and consumption of technology such as "Lifestyle," "Activities," "Interests," "Personality Type," and "Style." Lexical choices such as these all contribute to an understanding of technology within the ideology of choice (Kress and Adami 2010) in which as active consumers we can simply pick and choose those options on the menu that best suit us.

After the words have disappeared into the center of the screen, the line drawing of a futuristic city appears on the screen. At the bottom of the screen, the message "Stay tuned. . . . We're building your future" then appears. This is presumably the technological utopia that is being built out of your selections.

Once the allotted time has passed for the video to be compiled in accordance with the rider's supplied answers, the narration continues: "And now I believe your future is just about ready. Let's take a look, shall we?" At this point, the video is played upon the touch-screen (see Figure 7.1). The video narrator begins with "Welcome to the future, or should I say, your future!" The narrator proceeds to describe your future life growing vegetables with your riding companion, in this case, in the country in a large high-tech green energy house made of recycled materials with the faces of your riding companion and yourself imposed on the two animated figures in the video. Just as the questions were all about "you," the video is now all about "your": "your future," "your pedal-powered bike," "your green home," "your garden,"

DISCOURSES OF TECHNOLOGICAL (DIS)SATISFACTION 173

Figure 7.1 *Welcome to ~~the~~ your future!*

"your comfort," and "your living room." As the video ends, you are told that "You can live the good life and the green life in your home in the country because in the future, everyday is Earth Day." The future, your future, is tailor-made and the promises of the so-called green technologies ensure that you will be able to have your cake and eat it too. Thus, the descent down the geodesic sphere metaphorically brings the riders and technological innovation back to earth, grounding it in the "real" of the "personal." Even the animated short is itself a machine-generated personalized video.

Once riders have disembarked from the ride, they must depart through the Siemens branded post-show, which showcases Seimens R&D in biotechnology, medical technologies, energy management, and mobility. Entry into the after-show is marked by a sign that reads "Project Tomorrow: Inventing the wonders of the future Presented by Siemens," which leads to an open space divided into four interactive exhibit areas. "Body Builder—tools of the future using your genetic code" requires the player to assemble a digital 3-D animated skeleton and promotes Siemens' research into remote surgery and the development of robotics for tele-present surgery. "Super Driver—future of cars," and "Power City" are essentially driving games that promote Siemens' research into driver management and accident avoidance systems for motor vehicles. Players manage a power grid in "Power City—megacities of the future," which is the only multiplayer game and is a cross between shuffleboard and isometric perspective computer games like earlier versions of Sim-City. Finally, "InnerVision—predicting the future of medicine" involves the player in a series of memory and coordinated tasks in a promotion of Siemens' biotechnology research.

InnerVision in many ways exemplifies the notion of the future being a privatized individual experience. The game is designed to give the players an idea of what medicine will be like in their future. It invites the players to imagine they are at home doing what will become a routine daily self-maintenance health check in front of a high-tech bathroom mirror. One of the things that is common to all of the exhibits is the idea that the "impact" of these technologies is to be found in mundane, everyday settings and used and experienced as individual consumer-citizens. To play the game, the player performs a series of memory and eye-hand coordination tasks by pressing illuminated buttons that match with colored squares as they appear on the display. The upright display also includes a projection of a male body with the central nervous system illuminated. As the game proceeds, different parts of the nervous system are illuminated to seemingly correspond with what is supposed to be the inner state of the player. At the end of the game, three scores are displayed: Reaction Time, Memory Evaluation, and Hand-Eye Coordination along with the message "Sending Results to Doctor." What InnerVision presents is a vision of the future that is entirely in keeping with neoliberal health management regimes in which the individual not only takes on the responsibility for his or her own health management through self-monitoring technologies but also participates in his or her own surveillance through the sharing of this data. All this, of course, is made very convenient by embedding the technology in the home and integrating it into the daily domestic routine.

Me and my robot

In 2011, iRobot released a multimedia ad campaign in magazines, on television, and on the WWW to promote their consumer model robot, the Roomba. The campaign was called "iRobot. Do you?" so that the company name also functioned as a first person present tense speech act, offering information to the viewer in the form of a declaration. As it is explained on the Mullen agency website, "To raise awareness of their brand, we turned iRobot into a verb that was both inclusive and actionable" (mullen.com 2011). What the ads did, then, was try to make the company name also a slogan that represents Roomba ownership as something you "do." In effect, they sought to compound the nominalization, iRobot, with a process, thus playing on the notion that you are what you do.

In the print ads (see Figure 7.2), the corporate logo, iRobot, appears in large white letters at the center of the ad across the chest area of the photo model. Like all the lettering in the advertisement, it is contemporary sans-serif

Figure 7.2 *Bees. iRobot magazine ad.*

typeface. In this particular ad, "Bees," we see a woman holding up a Roomba robotic vacuum in front of her face so that it occupies the space in which her head would have appeared. Near the bottom of the ad, to the right, and in fine print, a kind of humorous testimonial is offered in which the woman explains how much she dislikes vacuuming (the problem) and that she owns a Roomba (the solution), and concludes with the word iRobot, which presumably is meant to be understood as a statement of how she has resolved her issues with vacuuming. All of the print ads follow the same layout, and include such a first person testimonial in the bottom right-hand side of the advertisement and conclude with iRobot alone on the last line. Below the testimonial and in a much larger green-colored typeface is the interrogative clause, "Do you?." Positioned as it is, the "Do you?" follows from the iRobot in the testimonial; however, given the small point size of the font, the question is more likely to be read in relation to the much larger and more salient iRobot at the center of the ad. If we consider the placement of text in relation to the information values, the centrally placed "iRobot" is clearly in a position of importance while the testimonial, located in the bottom right, is more marginally positioned. The right side is also typically the new side in given-new relations and while there is no text located to the left, the "Do you?" has the potential to be read as following from the centered "iRobot." The positioning of the Roomba robot in relation to the testimonial also has the potential to be understood in terms of an ideal-real relation with the Roomba as product occupying the position of ideal and the misery and animosity to the chore of vacuuming being the real.

While the advertising agency makes it clear that they want the corporate name iRobot to be understood as a material process clause, "I robot," with "robot" functioning as something the Roomba owner does, iRobot also clearly functions as a declarative of identity. Recalling the title of the collection of short stories from which the company name is derived, iRobot also has the potential to mean "I am robot." The placement of the large iRobot over the chest of the iRobot owner can also function as a label to identify the user not just as one who robots but also as one who is a robot. But it is the placement of the Roomba itself that is really suggestive of the notion that the user is declaring herself to be a robot. In each of the print ads, the user holds the Roomba in front of his or her face in the same way so that in each ad, the head of the owner appears to be a Roomba. It is through this visual cue combined with the word play of the advertisement that the popular abridged Cartesian quote "I think therefore I am" is evoked as "I robot, therefore I am robot."

The video commercial, "Robot Dance" (2011), from the same campaign also evokes the same idea of the owner performing and becoming robot. In the commercial, five dancers, including the inestimable Marie "Marie Poppins" Bonnevay and Marquese "Nonstop" Scott, are depicted as a supposed cross-section of different home owners, all of whom use a Roomba.

The video begins with a Roomba leaving the charging station and beginning a cleaning cycle. The same thing presumably happens in the other four households. The video then cuts to a Roomba moving across the floor of a designer's home studio space. The designer turns her head to look at the vacuum in a slow, smooth dimestop movement. The commercial then cuts to a man sitting on a couch reading a magazine in a mid-century modern-inspired living room. He stops and looks off screen, presumably at his Roomba, and then sits up in two dimestops. Next, a woman in the ornately paneled hallway of a "traditional" home, painted white, with a very high ceiling, sets down her briefcase as she looks at her Roomba. This is also done in a dimestop motion and she proceeds to do a robot dance in the mannequin style with a glazed, expressionless look on her face. The video then cuts to an older male who, in a loft-style bedroom, stands and straightens up in a dimestop with his forearms extended, hands out, and eyes looking upward. The camera then alternates

Figure 7.3 *Catching sight of the robot.*

between the five dancers robot-dancing in their respective residences. The commercial ends as Poppins abruptly breaks into a tight smile, resumes normal movements, and picks up her briefcase and departs off camera while the Roomba continues to vacuum.

The music for the commercial evokes a Latin rhumba style and the chorus sings "Roomba, Roomba" to the rhythm. There is a brief song sung in a raspy, high-vocal-tension-style by a female vocalist, which has a meaning potential of passion. The lyrics are somewhat inane but suggest great affection for the special vacuum in the singer's life. The final lyric is "Got a Roomba vacuum cleaning up my life!" The choice of the Latin rhythm, of course, has the meaning potential to suggest passion, and this redounds with the singing style of the vocalist. Allusions to the rhumba also add to the meaning potential of passion. A dance like the rumba entails dancers engaged in a seductive partnership. Interestingly, all of the Roomba owners are depicted alone, which means that the relationship between robot and owner is the sole focus of the commercial. It is when the owner looks down and sees her or his robot that he or she begins to dance in the robot-like style. When the owner does break into dance, it is as if he or she has entered a different and slower time-space outside the normal one. When the dancing stops, the owners return to their regular movements and speed of movement. The effect is that seeing the robot gives them a moment of stepping out of the ordinary time-space of regular life, and entering into another one occupied by the robot. Owner and robot come together and share this brief moment and then the owner ends the dance and resumes

Figure 7.4 *Experiencing a mini-sublime moment.*

the daily routine. It is as if each owner undergoes a small, mundane epiphany or sublime moment.

This is what Slack and Wise (2005: 19) characterize as the mini-sublime. They argue that the technological sublime with its ties to grand narratives of progress has been supplanted by what they call "mini-sublimes," which they characterize as the "cool" and the "neat." Seeing the Roomba cleaning your floors for you does not invoke a glimpse into the perfection of creation or make you part of a divine plan for the establishment of heaven on earth. Rather, the mini-sublime moment of the Roomba is just a passing aesthetic experience in which your personal world seems perfect for a brief instance. What the commercial does then is semiotize that brief moment of fleeting satisfaction when everything does seem convenient.

Robots failing miserably in the world of work

Slack and Wise (2005: 28) submit that our relationship with technology has increasingly been defined in terms of convenience rather than progress. As technology has been characterized as a means to the good life, the good life is increasingly to be found not in terms of where one stands in relation to progress but rather in the experience of convenience. As Slack and Wise (2005: 28) express it, "Convenience, more often than not, is the everyday motivation that justifies the ongoing choices involving the role of technology in everyday life." And yet, as Slack and Wise demonstrate, following from Cowan (1983), convenience is not necessarily convenient when you consider the broader set of interconnections that come with a particular convenience. Using the bread machine as their example, Slack and Wise (2005: 35) describe a number of ways the convenience of being able to keep fresh baked bread at home is not really all that convenient:

> This convenience, like all household conveniences, is part of a technological system that makes us more comfortable in some senses. However, the network of connections that constitute this technological system do not, in the end, reduce labor and save time; instead, the network of connections is part of a shifting burden in which the demands to collapse time (you can make that bread now!) and space (you can have that bread here!) become, in a sense, an inconvenience. These contemporary demands are burdens, responsibilities, and stresses that can only be called uncomfortable.

The bread machine was sold by promoting the idea that it was a convenient way to prepare fresh homemade bread for your family. But, of course, once it is on your counter, it will also need to have the ingredients that are processed to work with a bread machine, you will need to give over counter space to it, you will need to disassemble it, clean it, dust it, and keep the parts properly primed. In the end, the labor-saving device ends up delegating new duties and responsibilities to the householder. The result is that our search for satisfaction in technological conveniences often leads to dissatisfaction. Consumer culture, of course, depends upon our discovery of dissatisfaction in our quest for satisfaction, and so producers and marketers of consumer goods promote a "perpetual state of dissatisfaction" through the promise of convenience (Slack and Wise 2005: 40). The result is that we come to experience a sense of ambivalence toward our conveniences. On the one hand, they seem to promise the overcoming of some limitation or problem in our life, on the other, they never seem to quite work out as we imagined. Looking upon them as if they were black boxes, we fail to see them in the broader context of the technological systems to which they are interconnected.

This sense of ambivalence toward technology is exemplified in two commercials that use robots to humorously represent the world of work: *Robot* (2007) and *Marketing Cloud* (2012). On one level, they tell amusing stories about work and vulnerability in which the robot plays the hapless employee. On another level, by representing the robots as workers, they are represented not as capital but as labor and, therefore, as being capable of producing value. And through them we learn the precarious fate of value producers.

What particularly interests me are the ways in which the robots are being represented with the ability to take on the work of human workers engaged not only in productive, material forms of labor but in immaterial, affective forms of labor as well. It is as if robots, like human workers before them, are now moving out of the (post)Fordist assembly line factories and into what Mario Tronti (1966) has termed "the social factory." What the commercials in question depict, then, is the socialization of robotic "labor" or, more accurately, the representation of automation as a subsumed socialized labor. In sum, if robots once represented a form of work that was dull, repetitive, and individuated, they now seem to be increasingly represented in working environments characterized by agility, change, and teamwork.

In the 2007 GM commercial, *Robot*, the video opens on an automobile assembly line. Robots and humans are working on the semi-finished vehicles, installing parts. One robot drops a screw and makes an "uh-oh"-like sound, "looking" at the camera. A warning alarm goes off in the plant. The other robots stop and seem to look at the errant robot. Human workers also stop and look at the robot with annoyance and frustration. Slightly out of focus, a manager then comes over and places his hand upon the robot's "shoulder." At this point, the music has started. The robot turns his head back toward the manager and then drops it in resignation. The commercial then cuts to the manager and two line workers standing at the door of the plant. They nod to each other, shake their heads and begin to walk back into the plant. The video cuts to a medium shot of the robot leaving the grounds with its head down and the three plant workers returning to work in the background. The final shot of the sequence is an interior shot from within the plant showing a distant shot of the robot as the door is closed. The video then depicts the robot failing at marginal and precarious employment (sign waving and holding drive-thru order speaker). The video then uses a low-angle view of a bridge with the robot poised on the edge. The robot watches as all GM vehicles pass over the bridge and then jumps off the bridge in an apparent suicide. At this point, the robot suddenly awakes in the plant on the assembly line at night realizing it has all been a bad dream.

The second commercial, Adobe's 2012 *Marketing Cloud*, depicts a robot being called into the company Chief Marketing Officer's (CMO) office to be let go. In this commercial, the only action taking place is a dialogue, which includes an act of passive aggression on the part of the robot when it "accidently" drops its key card in the superior's coffee. Almost all the shots are medium close-ups focusing on the turn-taking and reactions in the dialogue. Following Hollywood narrative cinema conventions, the shots typically cut to the other interactant before the line is fully delivered so that the reaction of the listener can be established before the speaker takes on his role. The CMO is clearly uncomfortable with the task of terminating the robot and the robot does not make it easy for him. He tries to explain to the robot that they have found a better way to analyze marketing data and that it is "good on the math but not on the analytics." The robot even tries to save its own job by reporting another employee for taking long lunches and drinking at work. At this point, the CMO asks for the unhappy robot's key card, which it then drops into the superior's coffee cup. The commercial ends with the CMO holding up the dripping key card, looking back at the robot with an expression that might be interpreted as a blend of disapproval, annoyance, and disbelief.

As we have seen, our narratives about technology in general and automation in particular expose much in terms of our understanding of our relationship to the technologies that we develop and disseminate. This is a central part of Winner's (1977) argument in his highly influential *Autonomous Technology*. Thus, Tobias Higbie (2013: 100) has recently noted that in terms of primarily televisual and cinematic narratives entailing robots, we "reveal our ambivalence with the level of control we grant our technological mediators and anxieties about what it means to human amid pervasive technologies." Similarly, in her study of the constitution of "the posthuman" in advertising, Norah Campbell (2010: para 3) has charged:

> Like most other advertising images, technological images are not a neutral reflection of the world; technological images in advertising present future identities and worlds which are a crucial record of the contemporary social imaginary, revealing ideologies, fears, and fantasies accorded to technology.

However, Higbie (2013: 100) goes on to argue that in comparing apek's *R.U.R.* to more recent popular narratives of human-robot relations,

> one key difference remains between today's robot characters and those of the 1920s: workers. In their first incarnation, robots were clearly identified as workers and their theatrical rebellion with strikes and revolutions. Today, the experience and political economy of work, if they appear at all, are background details in robot stories, replaced by the thrill of shootouts between good-guy humans and bad-guy machines.

While this claim does seem to bear weight within the media that Higbie is addressing, in commercial advertising there have, in fact, been a number of fairly recent examples of representing the experience of work, although the representations typically centered on the robot's own "experience" of work. In the case of the two ads being discussed, it is precisely the experience and political economy of work as experienced by the robots that the humor hinges upon.

To borrow from John Law (Law 1991: 17):

> The very dividing line between those objects that we choose to call people and those we call machines is variable, negotiable, and tells us much about the rights, duties, responsibilities and failings of people as it does about those of machines.

Equally, the representation of robots as workers producing value through immaterial as well as material labor tells us much about the reorganization of work and the need of capital to subsume and expropriate value from the intrinsic social character of labor. What I am proposing, then, is that in the commercials we see a shifting in the boundaries between living labor and capital, between people and machines such that work is represented as a collective effort managed through interpersonal relations rather than bosses and, more broadly, it corresponds to broader systemic changes in the forms of labor-power workers are expected to provide in exchange for wages.

In my own place of work, I continue to note the myriad of ways that staff and faculty are called upon by the university administration to use their communicative and interpersonal skills to advance the branding of the institution and to increasingly participate in self-managing practices. As my university seeks to rebrand itself as "student-centric," faculty are pushed to cold-call and "warm" high-performing high school applicants to encourage them to choose our institution over others. In the name of student retention and success, there is increasing pressure on both faculty and staff to perform emotional labor in ways as trite as friendly sounding course outlines to proposals for untrained faculty to buddy-up with students who are struggling academically. Much like quality circles, the university has also established formal Communities of Practice in which faculty voluntarily meet to share and discuss best practice in teaching writing skills, using technology in the classroom, teaching large classes. At the same time, there is an increasing sense that teaching is reducible to a set of skills and performances that can be divorced from the "content" of what is actually being taught and that keeping students emotionally engaged and not bored is equal to learning. Finally, in terms of organizational planning, the university has implemented a resource planning and management exercise that is presented as a process that works laterally by using peer-to-peer information polling at all levels to define and enhance "areas of excellence." Ultimately, the goal is to pull information upward, which can then be used by administrative planners to justify the reallocation of resources. What these examples all illustrate are some of the ways in which sociability not only comes to be co-opted as an organizational or enterprise resource but furthermore is made obligatory.

What I wish to propose is that the humor in the way robots are being represented in the commercials lies in recognizing in their predicaments, the profound changes that have been brought about through the post-Fordist reorganization of work. Accordingly, in what follows I will identify and detail those semiotic resources that are regularly drawn upon in the commercials in order to represent agency on the part of the robots and to produce affectual relations between both the viewer and the robots and between the robots and other represented participants. Ideationally, I will argue that this is accomplished by the representation of the robots as social actors. Interpersonally, I contend that the endowment of the robots with the capacity for affective relations is established through the relationship between viewer and participant and relationships between participants. In short, the robots are represented in the commercials as possessing agency in their own right and also inner emotional states, which allow them to engage in affective relations with others around them. It is this blurring of boundaries between human and machine, labor and tool, and, ultimately, living labor and dead capital that tells us much about how we are being invited to reimagine the world of work and the rights, duties, responsibilities, and failings of workers in the new workplaces of compulsory participatory management.

Robots as agents

Not only are the robots represented as social actors in the two commercials but, in keeping with my broader thesis regarding the representation of robots as living labor, also as *sociable* actors. The allocation of agency is realized in the kinds of roles that are attributed to the represented actors. This highlights the degree to which there is no simple, direct relationship between the roles social actors perform in social practices and the (re)semiotized roles that they occupy within representations of the activity in question.

Both robots are represented as being *activated,* representing them as being engaged in material (holding a speaker, and roadside sign waving in *Robot* and dropping employee card in *Marketing Cloud*), mental (reflecting, longing, and feeling depressed in *Robot,* and sad and dejected in *Marketing Cloud*), behavioral (looking, and performing nonverbal cues such as facial expressions and emblems and speech-related gestures in both commercials), and verbal processes (the attribution of cognizance, and with it, agency, is further realized by the representation of verbal processes like the R2-D2-like "uh-oh" made by the assembly line robot after it drops the screw in *Robot* and the awkward verbal exchange between the robot being dismissed and its manager in *Marketing Cloud.* While obviously there is no claim being made that robotic technologies are really capable of such non-material processes, nonetheless, I will argue that the entertainment potential of these advertisements stems from an interest in viewing robots as autonomous agents and in recognizing ourselves in the predicaments of the robots.

At the same time, while robots can be said to take on activated roles in the commercials, humans, on the other hand, tend to be represented in *passivated* roles whereby actors are presented as undergoing, or at the receiving end of, the represented activity. In the two commercials, humans take on largely reactive and therefore passivated roles in *Robot* and *Marketing Cloud* where, in the former, the only action initiated by a human is when the line manager discharges the robot after it has dropped the screw and the other human and robot line workers react with nonverbal expressions of disapproval and impatience and, in the latter, we see a senior manager severing the employment of a robot whose negative reaction to the dismissal creates acute discomfort for the manager.

What I would propose, therefore, is that the commercials realize what I have already argued is a general tendency to present robots operating independently and autonomously as if they are self-determining entities. In each of the commercials, the robots are able to function without direct human intervention and when humans are represented, their interactions with the robots tend to be as either reactors to or as recipients of robot actions. In this way, robots seem to have an agency of their own, able to effect change to the material world, but also able to think and behave as independent living beings. In short, they too may be counted among the living that labor.

Robots as affective bodies

Any text is always more than an act of representation since it also functions as an act of exchange. So while ideationally the robots are represented as endowed with autonomy and agency, I would charge that the attribution of agency to the robots is further established interpersonally through the (re)semiotization of affective relations between the viewer and the robots and between the represented participants within the commercials.

Affect is closely tied to agency. Brian Massumi's (Deleuze et al. 2014: xvi) proposal that affect, or *L'affect*, implies "an augmentation or diminution in that body's capacity to act" makes this linkage with agency explicit. At the same time, affection, or *L'affection*, Massumi (Deleuze et al. 2014: xvi) tells us, supposes "an encounter between the affected body and a second, affecting, body" and therefore situates affect in the interpersonal or, more properly, the pre-personal. And while neither affect nor affection is synonymous with feeling or sentiment, the use of emotive elements, nonetheless, provides a means of establishing mood and the attitudinal orientation of the viewer to the robot in the given commercial as well as endowing the robots with the capacity for inner emotional lives. In this way, emotive elements can be understood to directly contribute to the realization of semiotized affective relations.

In this section, I attend to those semiotic resources that are implicated in the realization of affect within the interpersonal function of the commercials.

More specifically, I am focusing on how the organization of the participant relations in the commercials constitutes the robots as subjects capable of affective relations. First, the orientation of the viewer to the robots as represented participants is considered. Then, I review how the robots utilize nonverbal communication cues and are therefore seemingly capable of communicating interpersonally as well as representationally. Finally, I address the use of music in the commercials to provide further affective cues regarding participant relations.

Both commercials make use of demand gaze. In *Robot,* it is used in two key moments: when the robot appears to face the viewer both when it makes the error of dropping the screw thus holding up production and when it suddenly awakes in a darkened assembly plant and realizes much to its apparent relief that the error and everything that transpired after it was all an anxiety dream. In the other parts of the video, offer gaze is used, which lends itself to the general sense of despair and isolation that hangs over the rejected robot. Demand gaze is also used in *Marketing Cloud* where the viewer's perspective is that of the manager who is trying unsuccessfully to manage the interaction between the robot and himself as he dismisses it for unsatisfactory performance.

Vertical angle is used in an interesting manner in *Marketing Cloud* where the robot being fired tends to be shot at a lowered angle while the manager is often shot at a slightly raised angle. This has the effect of adding to the awkwardness and humor of the situation. It is only in the final shot of the robot, once it has deliberately dropped its key card in the manager's coffee cup and the exchange is over that the robot is shot from a higher vertical angle. What such uses of vertical angle suggest is that while the robots may not be elevated in status as our overlords, they are, nonetheless, being represented as significant and worth viewer attention.

Just as for gaze and angle of interaction, the distance between viewer and robot participants also tends to be dynamic. In *Robot,* intimate, social, and impersonal distance are all used to effect and intimate shots are used typically when the emotional state of the robot is being conveyed. Intimate social distance is consistently used in *Marketing Cloud* and this is typical in terms of the conventions of filming a dialogue. Overall, impersonal distance is rarely used and this is in part, I believe, because the robots are by and large being presented as being capable of engaging in interpersonal relationships.

What I would argue, therefore, is that the positioning of the viewer in relation to the robots contributes to their constitution as bodies with the capacity to form affective relations with others. While the viewer may not be called upon to directly engage with the robots, he or she is typically oriented toward some degree of involvement, and positioned socially and occasionally intimately, with the robots. What this suggests is that the robots are being presented interpersonally as subjects rather than objects and therefore are social beings themselves with the potential to reciprocate viewer involvement.

If the viewer orientation implies that the robots represented in the commercials have the potential for forming interpersonal relationships, the use of nonverbal communication by the robots makes this explicit. It is quite apparent that the robots do not simply employ the "emotionless" synthesized speech of earlier science fiction robots but, instead, incorporate a wide range of nonverbal cues. This is because their communicative acts are not simply representationally but also interpersonally oriented.

Of course, the robots essentially have fixed facial expressions but even in the absence of dynamic facial expressions, they are able to convey inner emotional states using gestures and posture. The robot in GM's *Robot* has two screws positioned where we would imagine eyes and a beaklike caliper that completes its expressionless face. Nevertheless, it is able to convey its inner emotional state through posture by slightly turning and lowering its "head." *Marketing Cloud*, in turn, offers a more sophisticated rendering of nonverbal communication since the entire commercial is a difficult face-to-face conversation. Here the robot uses speech-related gestures, posture, eye behavior, and vocal behavior to reveal its unhappiness with the situation.

Finally, the use of music in *Robot* is clearly implicated in the creation of mood and, therefore, affect. The robot featured in the commercial can be described as "minimally expressive." Minimally expressive robots are used in some forms of autism therapy to presumably teach children generalizable "rules" regarding emotional expression without overloading them with too much information, the idea being that fewer cues are less ambiguous and, therefore, more readily understood. In the case of *Robot*, the limitation in expressiveness in combination with other elements, such as music, actually works to overdetermine the emotional states of the robot for the viewer. Thus, the musical score not only provides cues to orient the viewer in relation to the robot but, particularly in the case of non-diegetic music, also provides cues as to how to interpret the inner emotional state of the robot.

This is exemplified in the musical score selected for *Robot*. The music is delayed, making the music clearly redound with the emotional state of the robot. In the commercial, Eric Carmen's *All By Myself* begins to play only at the moment the line manager walks up and places his hand upon its "shoulder." The song starts at the instrumental section between the first and second stanza and continues to the first verse of the third stanza: "Livin' alone. . . . All by myself." The selection of lyrics, of course, highlights the social exclusion that the robot faces upon committing its error and being dismissed from the assembly line.

Figure 7.5 *Leaning forward with "eyebrows" upturned, the robot receives bad news about its employment.*

Turning to semiotic resources in sound and sound quality (Machin 2010: 120–26) in the selected section of *All By Myself*, we see that the use of music to both orient the viewer and convey the inner state of the robot is all the more apparent. The vocals are sung at a lower pitch realizing a certain heaviness and despondency, but it ascends with the verse "All by myself," emphasizing the feelings of loss and isolation. Carmen's vocal qualities realize a heightened tension along with some audible raspiness and nasality, which I would propose suggests intensity of feelings. Likewise, the presence of vibrato further realizes the experience of emotional suffering. The melody is slow and low in key, but also what Machin (2010: 107) characterizes as "untroubled" such that "it communicates something of the singer's mood. There is no complexity, trouble or doubt in what he feels." Thirdly, the phrasing of the melody entails a slow fall or release followed by a more rapid rise; then returning to a fall at "self" creates a kind of lingering and longing of emotion that is punctuated by an outburst of self expression of solitude but then it begins to return to the previous sustained emotional tension. The result is a highly melodramatic form of pathos.

By orienting the viewer in relation to the robots such that interactional relationships between viewer and robot are constituted, by representing robots as using nonverbal cues in their behaviors and interactions, and through the use of music scores that offer additional emotive information, the robots take on emotional lives of their own. What this suggests is that the robots are being presented interpersonally as subjects rather than objects and therefore are social beings themselves with the potential to reciprocate viewer involvement. Accordingly, it is by representing robot bodies as affected and affecting bodies with the capacity to act that robots are made to come alive with technology and, therefore, become a kind of living labor. Moreover,

Figure 7.6 *Assembly line robot is dismissed from the team.*

the processes performed by robots, by implication, are themselves social labor rather than simply mechanical force.

Do robots dream of precarious labor?

For Maurizio Lazzarato, commencing roughly in the 1970s, work begins to be restructured such that the division between mental and manual labor becomes increasingly tenuous. Lazzarato (1996: 134) argues that "a new 'mass intellectuality' has come into being" such that an uneasy synthesis has been arrived at between, on the one hand, the requirements of post-Fordist forms of production for a workforce that is knowledgeable, adaptive, and able to quickly choose appropriate courses of action and, on the other, the demands by workers seeking "self-valorization." The result is that capital has sought to re-absorb the struggle against work as a new form of control in the labor process, which Lazzarato (1996: 134) characterizes as "the process of valorization." The point is not so much that the labor process has appropriated intellectual or immaterial labor but that capital paradoxically seeks to make workers intellectualize their means of subjugation by making them take part in their own management. The innovation of participatory team-forms of management is, therefore, in point of fact, part of a broader set of innovative social technologies for producing subjectivities that are amenable to what can be characterized as, following from Lazzarato (1996: 134), the new workspaces of control.

This restructuring of work and the corresponding encroachment upon the affective-subjective lives of workers has, therefore, meant the transformation from a disciplined proletarian workforce to a "labour of control" (Lazzarato 1996: 134) as capital now extracts value from post-disciplinary socialized rather than an individuated labor. Whereas, previously under Fordism, the production cycle was primarily accomplished through the disciplinary administration of a workforce such that the toil of many individual workers could be accumulated and value extracted, within the "postindustrial" economy, the reorganized cycle of production is no longer confined to within the walls of the workplace and, instead, takes form as and when needed from the broader social networks and flows that make up what Lazzarato (1996: 137) refers to as "the basin of immaterial labor." The success of "postindustrial" enterprises, therefore, increasingly depends upon the ability to spontaneously draw from and reorganize the productive immaterial *and* material forces of the *social* body rather than just the ability to harness the direct or immediate labor of massed individual workers' bodies.

In keeping with this restructuration of work, what the commercials do is present the robots as having undergone a parallel transformation such that they are implicated in affective and immaterial as well as material forms of labor. In *R.U.R.* robots are "working machines":

> But a working machine must not play the piano, must not feel happy, must not do a whole lot of other things. A gasoline motor must not have tassels or ornaments, Miss Glory. And to manufacture artificial workers is the same thing as to manufacture gasoline motors. (Capek and Selver 1923: 9)

But in the commercials, the robots are, instead, *social machines*. In other words, if the robots of *R.U.R.* took on the work of the proletarian labor of the 1920s, then the contemporary robots of the commercials are ready to take on the work of socialized labor. They have literally moved from the factory floor to the social factory.

Referring to actual automated machinery, Caffentzis (1997: 41) asserts that in accordance with the Labor Theory of Value, "these automatic machines simulate the role of a worker's concrete useful labor but they cannot create value as the worker can by actualizing his/her labor power into abstract labor." The fictional robots of the commercials seemingly overcome this limitation.

Within the Labor Theory of Value, the labor process, according to Marx (2005a), comprises three primary elements: "1) the personal activity of man [sic], *i.e.*, work itself, 2) the subject of that work, and 3) its instruments." As Caffentzis (1997: 33) has commented, capitalism, in seeking to exploit labor and extract surplus value actually disintegrates the labor process through a series of separations:

> First, primitive accumulation detaches the laborer from the subject of the labor, and then manufacture, through its obsessive attention to the details of the "naturally developed differentiation of trades", separates the work of the laborer from the instruments of that work to the point where Modern Industry appears to do away with the *activity* of the laborer all together.

In short, the robots become a kind of simulacrum of living labor and so offer the possibility of deferring the acknowledgment of the final disintegration of labor process under an imagined total automation.

In representing robots in this manner, the commercials reveal much in terms of our understanding of our relationship to the technologies that we develop and disseminate. In other words, my concern here is not to simply reveal misunderstandings of what the technologies can or cannot do but rather to address how robots have come to both represent the new future of work and forestall thinking about what that future of work might mean.

Accordingly, by presenting the robots as actors with agency that participate in affective relations, they become more than simply autonomous technologies; they are, instead, autonomous agents. In this way, the robots are being represented not just as being capable of laboring as humans but also as equally capable of performing the cognitive and sociable forms of labor appropriated into the post-Fordist organization of work. Quite simply, the robots do not simply imitate human labor but rather simulate it (cf. Baudrillard 1994) by seemingly having the capacity to work socially as well as physiologically. At the same time, as robots, they lack the capacity to refuse work and, therefore, equally, the capacity to refuse to be sociable. Thus, they appear to simulate not just concrete or material useful labor but immaterial useful labor too, and thus seemingly cease to be instruments, coming to life so that dead labor may truly suck the life of living labor. Ultimately, what the commercials offer are entertaining narratives that transpose the real world of work into one that seems more laughable. Following from John B. Thompson (1984: 6), they are part of the broader "creative, imaginative activities [that] serve to sustain social relations which are asymmetrical with regard to the organization of power." They allow the viewer to see their own conditions of employment and still laugh since the tables have turned and now it is a robot that is losing its job.

Conclusion

The discourse of technological (dis)satisfaction represents a more modest claim about technology and the future it promises. Rather than promising salvation for the chosen people as we see in the progress narrative, it simply promises increased convenience to the user. Technological innovation becomes a matter of what it will do immediately to improve everyday life by overcoming some problem or limitation. Few people buy robot vacuums as an expression of their patriotic duty and national salvation. The automated vacuum robot simply promises to clean floors without your having to do the labor yourself. Of course, technologies such as these are only capable of doing what the design actually affords. So in one case, a robot vacuum cleaner owner reported an incident in which the robot performed its cleaning cycle as it was supposed to but the family pet had opened its bowels that morning on the living room floor. The robot, programmed to move throughout the room, spinning its cleaning brush, proceeded to spread the dog excrement throughout the room.

The shift to understanding technology as being a matter of convenience also results in an experience of dissatisfaction. New technologies are marketed to be convenient and the experience of seeing the possibility of convenience is like a small, personal sublime moment. The problem is that if what we have truly gave us satisfaction, we would not have need for more. Capitalism depends upon us needing more. In terms of new markets, unlimited expansion is simply not possible and yet capitalism presumes unlimited expansion. In lieu of unlimited expansion, marketers and producers seek, instead, to promote unlimited consumption. The result is that like built-in obsolescence, consumer culture depends upon built-in dissatisfaction. The experience of convenience is fleeting and so the return of dissatisfaction ensures the continued pursuit of convenience. Thus, embedded in satisfactions are dissatisfactions. The result is that our relationship with technology is marked more by ambivalence than it is in the cliched techno-optimism of the mid-twentieth-century.

Conclusion

I have argued that, in contrast to commonsensical understandings of technology, technologies should not be thought of as neutral, transpicuous tools and that they are, in fact, confluences of knowledges, activities, and materials that extend beyond the immediate physical limits of the device. Technical objects are organized, invested with significance, endowed with values and capacities, and incorporated into social action and relations, and so our discourses on technology are always realized through the interconnection of signification, material artifacts, techniques, and cultural values. To this extent, I contend that the concept of apparatus better serves us in understanding the politics of technology.

An apparatus creates durable relations between elements so as to produce a disposition of gestures, behaviors, opinions, and discourses, which is embodied in the technology-subject relation. A technological object, therefore, never acts alone upon the subject but, rather, always enters into a relation with the other elements assembled within the apparatus. As heterogeneous ensembles, apparatuses have effects and consequences, they make a difference, and they matter. The politics of the technology-apparatus rests in its "capacity to capture, orient, determine, intercept, model, control, or secure the gestures, behaviours, opinions, or discourses of living beings" (Agamben 2009: 14).

The four discourses on technology, as progress, technological determinism, technological fetishism, and technological (dis)satisfaction, are all closely related. The discourse of technology as progress placed technology on the center stage and made it a yardstick of civilization. Progress could literally be seen in the industrialization of the landscape and it was the machine that now brought providence or at least was a sign of being among the elect. The discourse of technology as progress meant that technology was not only a sign of divine intention but also that technology had the power to transform nature into "man's" dominion. Although in "American Progress," technology and progress have not entirely merged and technological innovation is more the harbinger of progress than progress itself, increasingly technology itself is becoming progress. So while in Gast's painting of "American Progress," progress is depicted as a blessing bestowed upon select nations, as we see in the Carousel of Progress, it increasingly came to be bestowed upon individuals. Progress was not just something that could be experienced abstractly in terms of national spirit, but progress increasingly found its way into the private home. The intersection of the Cold War with the rise of post-Second World War consumer culture meant that if you had the money, you could buy and bring home a bit of progress yourself. Progress came to be democratized as something one could be a part of through purchase of the right things. Accordingly, as the Carousel turns through each period, we "learn how the technological marvels of the day have made life more convenient, comfortable and fun" (Disney 2015). At the same time, as Noble (1984) points out, not everyone necessarily believed in the promises of progress and thus some did express ambivalence and dissatisfaction over the signs of technological progress. As such, the discourse of technology as progress can be said to have opened the door to the other three discourses on technology that have been discussed in this book.

Technological determinism continues the logic of privileging technology not just as a harbinger of progress but as the primary motor of social change. However, technological determinism also departs from the teleological bent of the progress narrative. The discourse of technological determinism makes technology the underlying base of a society's development, but it does not intrinsically assume that the process of change is directed toward betterment. Technological determinism offers an explanation of why a society is the way it is by making technology the definitive core or essence of that society and then tracing its effects through the society. Of course, a given technology could be as much a source of negative effects as it could be of positive effects. In the "Look Up" (Turk 2014) video, mobile devices and social media are determined to be destroying our ability to communicate face to face. Because technological determinism privileges technology as being responsible for the social condition, it is open to praise or blame depending upon our feelings toward that set of conditions. One thing technological determinism does clearly retain from the progress discourse is the idea that technologies develop in a linear, predictable way independently of social, economic, and political conditions. Technologies are construed as developing causa sui.

I have argued that technological fetishism represents an intensification of technological determinism. While technological determinism does not intrinsically make claims about the "value" of a technology, technological fetishism depends upon the hyper-investment of social value in the fetish object. Both technological fetishism and technological determinism privilege the technology as having a special capacity to produce effects upon society, but a determining technology does not have to possess fetish value. Indeed, in terms of how fetishism was redefined not simply as the misattribution of special powers but rather as a sign of social labor, the mobile device

CONCLUSION

in "Look Up" was, to Turk, not so much a fetish as it was a scapegoat. Technological fetishism, like progress and determinism, treats technologies as if they are discrete and autonomous things able to act upon the world while disassociated from the broader set of connections that situate the technical object within a technical system or what I have been referring to as an apparatus.

The discourse of technological (dis)satisfaction reflects the shifting emphasis upon technology from a sign of progress to a matter of convenience. It represents a resetting of the stage upon which technology is made to perform in these discourses. In the discourse of progress, technology performed in a grand narrative about the exceptionalism of the nation-state while in the discourse of technological (dis)satisfaction, technology is, instead, made to perform in more mundane, everyday settings. As technologies are promoted as being more tailored and personalized, our expectations also rise. When consumer technologies were promoted through the discourse of progress, they were promoted as being good for everybody. These were mass-marketed technologies as opposed to the more niche-oriented marketing of contemporary post-Fordist consumer culture. It meant that they did not have to suit us perfectly. Selling goods and services on the basis of convenience, instead, means selling on the promise of solving our immediate personal experience of dissatisfaction. Since the experience of convenience is a fleeting one, dissatisfaction quickly returns. In fact, consumption depends upon the return of dissatisfaction and so it is immanent to the pursuit of satisfaction and convenience under consumer capitalism.

All of these discourses share in the same conceptualization of technology. Constituted as discrete, finite things that are disassociated from the milieu to which they have become associated, they limit our experience of technology to that of the black box. This is epitomized in an article on industrial design appearing in the Siemens corporate magazine, *Pictures of the Future*, where the author writes, "New products make it in the marketplace only if their complexity is invisible and customers can use them quickly and intuitively" (Weber 2012: 58). In other words, the four discourses I have sought to describe and critique all depend on this same theory of technology as reducible to neutral, isolated, useful things. In problematizing this theory of technology and supplanting it with a more radical one, we can begin the work of articulating a very different discourse on technology, one that understands that it is profoundly political and deeply implicated in the technocultures in which we live.

Mackenzie notes that within the Social Construction of Technology (SCOT) literature, there has been a tendency to equally by-pass considerations of the role that more enthusiastic promotional discourses, or quite simply "hype," might play in constituting the technical object: "The cost has been an immunity to the idea that the hype might be constitutively part of technology as we know it, that hype can get under the skin, so to speak, of technologies, whatever their materiality or sociality" (Mackenzie 2002: 383). He attributes this, in part, to SCOT's rejection of deterministic, humanist accounts of technology as a kind of negating force. Alternatively, Mackenize, argues that attending to any hype that might surround the technical object affords consideration of affect, which he finds lacking in SCOT accounts of technology. As he continues, hype "does connote an excess of feeling or affect, an over-attachment to an image or figure, usually circulating through mass media or publicity mechanisms" (Mackenzie 2002: 383).

Mackenzie (2002: 393) proposes that Simondon's conception of transduction foregrounds the equally constitutive role that affective relations play in the individuation of the technical object since it "extends to perception, feeling, acting, knowing and thinking as ontogenetic processes." If we forge relations with and through technology then this is accomplished affectively as much as it is socially. In this way, "collectives can come into being around technical activity" (Mackenzie 2002: 394) and it is in this way, therefore, that technology, as a materialization of sociality, is inherently political.

While Mackenzie refers to hype, I believe it is fair to extend this beyond enthusiastic talk about technology to discourses of technology as a whole. What I have sought to do in this book is identify some of the primary ways semiotic resources are drawn upon to realize the discourses of progress, technological determinism, technological fetishism and technological (dis)satisfaction that frame received understandings and experiences of technology. In adopting a Simondonian approach to the (re)presentation of technical objects, my intent has been to argue that the meaningfulness of technologies is not given or self-evident, but at the same time, technologies should not be understood as tabula rasa either. We cannot resolve technological determinism by merely supplanting it with cultural determinism and it is a mistake to assume that meaning just "sticks" to objects. Socio-technical relations are always expressly political since "relations to others, to machines or technical ensembles, and relations to self are knotted together in Simondon's account through processes of individuation" (Mackenzie 2005: 395).

Therefore, when we attend to the process of resemiotization and the generation of meaning-materiality complexes, we are, in effect, conceptualizing semiosis as an emergent, heterogeneous, relational phenomenon. We are, in fact, not so far off from what Akrich and Latour (1992: 259) characterize as semiotics in that we are not treating meaning as reducible to signs. According to Iedema (2003: 31), the multimodal turn in social semiotics has been facilitated in part because, in keeping with Halliday's systemic-functional linguistics, "semiosis [is] not analysed in terms of discrete building blocks or structures, but in terms of socially meaningful tensions and oppositions which could be instantiated in one or more (structural) ways." If semiosis is already understood as the product of relations and multimodality has increasingly stressed the heterogeneity of those relations at the level of the text, then it should not be such a great leap to begin to think of meaning-making transductively. Consequently, since "discourse" is a constitutive element of the technical object itself (not just what can be "known" about it but, instead, how the object comes into "being"), then a CDS approach is ideally suited to critically analyzing the individuation of the technical object and, ultimately, the kinds of relationality in which it is implicated.

In the end, I am proposing a CDS approach to the study of technology because I believe it affords us a more interdisciplinary approach to the study of discourse. As a broad, eclectic but nonetheless rigorous approach to the social production of meaning, CDS regards semiosis as a material practice that is generative of meaning-materiality complexes. In this way, technology cannot be reduced to a blank slate upon which culture can be written (cultural determinism), nor is it an imposing force that leaves its mark upon the culture it touches (technological determinism). By understanding technology to be always already interconnected to broader technological systems, we can begin to talk about the significance of technology without overprivileging it and assuming it to be an autonomous force.

Notes

Chapter 1

1 To be fair to the research team, they reported that a male equivalent had been considered but that the location of male undergarments were too far from the heart to be effective.

Chapter 3

1 Discussion of these semiotic resources is by necessity terse and so I would encourage the reader to examine the literature cited in this section of the chapter for an expanded discussion.

2 I have omitted "automatically" here, which is part of Latour's hyperbole. His phrasing suggests that any explanations grounded in considerations of power, society, and discourse are done in a kind of knee-jerk, unreflexive, and, therefore, uncritical fashion.

Glossary

Affordance: The concept of affordance is taken from the work of James Gibson where he characterizes them as latent action possibilities. Both Kress and van Leeuwen apply this concept to semiotics in order to describe the meaning potential latent in semiotic modes and resources. Semiotic resources, for example, do not simply bear an associated meaning but, rather, also bring with them undeveloped but nonetheless potential meanings that have yet to be brought into social significance. In social semiotics, the theory of affordance is used to highlight the way resources differ from signs in Parisian semiotics.

Apparatus: An apparatus is a kind of material-ideological structuring of knowledge and action. It is variously theorized by Louis Althusser, Michel Foucault (as dispositif), Gille Deleuze (as assemblage), and Giorgio Agamben, among others. In this book I am borrowing from Agamben's definition to describe a heterogeneous collocation of discourses, practices, settings, and material objects that operate together to produce certain socially needed and enduring subjectivities and techniques. Defining technology as apparatus moves us beyond thinking of it as self-evident finite objects or tools that merely serve practical purposes.

Associated Milieu: Gilbert Simondon proposes that every technical object evolves and adapts in relation to its own environment or milieu. Every technical object, therefore, has its own associated environment or milieu in which it functions and different milieus will play host to differently adapted technical objects. Object and milieu are mutually determining. Like the technical object itself, the milieu is also always becoming. Changes in the object will necessitate changes in the milieu in which it operates just as changes to the milieu will bring about new adaptations to the object. In keeping with Simondon's emphasis upon becoming rather than a finalized state of being, this is understood as an ongoing process such that as the object is evolved, the milieu in which it functions comes to be (re)associated with the object.

Concretization: Simondon uses this term to refer to the process through which the becoming technical object arrives at a provisional stability.

The concretization of the technical object is not meant to refer to the arrival at a final material form but, instead, that as an object-in-process, the technical object achieves a durable but nonetheless contingent state—and always in relation to its associated milieu.

Conduit metaphor: Instrumental conceptualizations of language and communication frequently draw upon the conduit metaphor as a way of understanding how meaning is shared between communicators. From this perspective, language and other modes of communication are understood as channels or pipelines through which information is carried. This presumes that there is an ideal state in which communication can take place whereby information is transferred through the conduit unimpeded. From this perspective, language itself is just a carrier of mental content and does not affect or shape the meanings that it is used to carry.

Connotation: This term is used to describe the "higher-order" conceptual or symbolic meanings that are associated with a given semiotic resource. For Barthes, a text bore two levels of message, the more direct, literal, or referential meaning termed the denotative level and the more figurative or symbolic level in which ideology is manifested. Connotation, in Barthes work, can be understood to be the ideological evocation of tacit assumptions and meanings. For example, an image of a closed hand denotes a fist, but depending upon context, it can connote such concepts as strength, resistance to oppression, aggression, assertion of power, and so on.

Discourse: This term has been used in a number of different ways from simply referring to a verbal exchange to systems of representation that operate inter-textually and inter-discursively. In CDS, discourses are understood as functioning as socially constructed resources for representing some aspect of reality. However, discourses are more than just held ideas or beliefs since they are always tied to specific interests and guide how we interpret, act out, and reproduce the social world. This means that discourses are not simply a *product* of the social world and they, in fact, contribute to the reproduction of social relations by constituting participants, actions, manners, times, spaces, and so on. As such, discourses are realized in communicative performances or events through the selection and articulation of semiotic resources. Discourses are then analyzed on the basis of their being able to establish readily recognized conventionalized patternings of semiotic resource options within and between texts.

Hylomorphism: Aristotle conceived the relationship between matter and form such that matter is relative to the form that it takes. The genesis of the individual is therefore the product of a meeting between pre-existing form and matter. Clay is relative to the brick because the brick is made of clay piped into a brick-shaped mold. This is essentially a dualist model since the brick form is treated as a product of a meeting between the unformed clay on the one hand and a mold on the other. It is the brick mold that determines the final shape that the raw clay will take in order to form a proper brick. A hylomorphic theory of the relationship between technology and culture equally treats the two as pre-existing one another but also in relation to one another. Thus, a technological determinist perspective would assume that culture is formed in relation to technology while a cultural determinist position would reverse matters and treat technology as being formed in relation to culture.

Image acts: Halliday proposes four distinct categories of speech act or interact: (1) demand information, (2) demand goods and services, (3) offer information, and (4) offer goods and services. Images, by contrast, cannot realize the same four functions in terms of distinguishing between goods and services and information. Instead, in realizing an imaginary relationship between viewer and represented participant, images are categorized in terms of two interacts or image acts—demand and gaze. Demand images are images where the represented participant is facing the viewer and thus establishes an imaginary relationship in which something is being asked or demanded of the viewer. Offer images, in contrast, are images in which the represented participant is not facing the viewer and so is offered up to the viewer more as an object of contemplation.

Individuation: In contrast to individualization, individuation implies a process that does not lead to a final and complete state of development. Instead, individuation suggests a process (transduction) in which the individuated being is always in a state of becoming though it may well function in a semi-durable or metastable form. While individualization works toward completeness, individuation is a never-ending ontological process of becoming since the individuated subject always retains some element of the pre-individual and, therefore, the potential to become.

Logocentrism: The idea that communication, in its most fundamental and purest form, is linguistic communication and thus the most immediate manifestation of thought.

Metafunction: From Halliday, we understand that every semiotic mode must be able to afford its users the ability to accomplish three simultaneous meaning-functions or tasks: (1) producing representations of social action, (2) creating interactions and constituting relations between social actors, and (3) realizing those representations and interactions through the composition of specific kinds of texts. Accordingly, all communicative events entail the realization of three corresponding metafunctions: the ideational or representational, the interpersonal or interactional, and the textual or compositional. Communication can therefore never be simply reduced to an act of transmitting information; it is always bound up in the ongoing constitution of social relations and organization of semiotic resources.

Metaphor: A metaphor *transfers* a concept, idea, or quality from one thing (source domain) to another (target domain), while a connotation differs insomuch as it *associates* a concept, idea, or quality in conjunction with a denotative meaning. In this way, a metaphor can be said to be connotative but not that connotations are metaphorical. Metaphors afford the expression of experience by drawing upon other "like" experiences and in so doing highlight some aspects of that experience while downplaying other aspects. In this way, metaphors draw upon concrete experience but also shape how we interpret those experiences.

Metastable: Rather than treating phenomena as finalized forms, the concept of metastability suggests that within dynamic systems, those things we treat as having achieved stasis are in actuality semi-durable and mutable and accomplish only a provisional stability or stasis. In social semiotics, the concept of metastability is largely introduced by Jay Lemke (1993) to describe the relationship between social and semiotic systems.

Modality: This is the social semiotic approach to truth. What is of issue is not how true a representation might be but rather how true it is represented to be. For this reason, modality is understood to be an interactional resource rather than a representational one. Implicit in modality is the relationship between communicators as it is realized through the kind of claim to authority and truth being made. Simply put, modality is the degree of truth as the interactants understand it and the resources they opt to use in order to represent it.

Multimodality: This is an approach to communication in which meaning is understood to be accomplished through the interaction of semiotic systems or modes rather than through single modes. Historically, linguistics has privileged language as the dominant mode of communication and treated other modes such as gesture, attire, imagery, and spatiality as subordinate and supplementary. Multimodality stresses that meaning is an emergent property of modal interaction and not simply the sum of the meanings realized in the different modes.

Relationality: This term refers to the reality of relations and marks a growing interest in the exploration of the relational ontologies. It draws attention to how the subject and reality are constituted through relations. In Simondon, technological objects are not defined in terms of intrinsic properties but, rather, are concretized only insofar as they can be individuated and this is always accomplished relationally. The process of individuation produces a "be-ing" and this semi-durable or metastable state of being is always constituted through relations; as Del Lucchese (2009: 181) succinctly puts it, "Being is not what 'is'. . . . Being is what becomes in and through relationality."

Resemiotization: Inspired by the Actor-Network Theory concept of translation, Rick Iedema (2001, 2003) offers this term to refer to the process of retextualizing and recontextualizing from one semiotic mode to another. Translation is not a neutral process since meaning can never be neatly transferred from one language to another. There is always something that is lost and also something that is added. The concept of resemiotization, accordingly, affords consideration of not only how semiotics are translated from one mode to another but also why a particular semiotic is enlisted in a given context and what semiotic work has been delegated to it.

Semiotic mode: This is a special category of "socially shaped and culturally given" (Kress 2009: 54) resource that social actors regularly draw upon for communication. Indeed, the principal function of any semiotic mode is communication. Each mode affords communicators the ability to accomplish meaning making across all three metafuntions (representational, interactional, and compositional) by making available semiotic resources that are specific to that mode. Some obvious examples of semiotic modes would be speech, images, music, material culture, and typography. Each of these functions has a means of communicating, with its own potential for meaning making, and has come to be used in regularized ways within specific socio-cultural groups.

Semiotic Resource: Rejecting the concept of sign as it is theorized in Parisian semiotics, social semiotics refers to semiotic resources instead. Similar to signs, semiotic resources are signifying elements or options that are available to us in order to compose and interpret texts. The fundamental difference, however, is that the concept of resource affords greater consideration of the diachronic features of signifying practices. This is because, unlike signs, semiotic resources have meaning *potentials* that are then determined *in use* and *in relation* to the other resources selected by the communicators.

Semiotic resources, therefore, have a range of potential meanings in contrast to a clearly defined referential meaning and the meaning potential of semiotic resources rests, then, both in convention and in what current circumstances can afford.

Technical lineage: The development of a technical object always occurs in relation to its associated milieu through an ongoing process of individuation. Rather than understanding this as a linear process of finalization, Simondon, instead, proposes that technical objects undergo evolutionary processes much like organisms do in relation to the affordances of their own environments. In this way, the evolution of technical objects can be traced through genealogical lines or technical lineages. The concept of technical lineage thus bypasses the tendency to frame the development of technologies in terms of continuity, linear causality, and teleology. In contrast to a process of individualization, which leads to a predetermined state of completion like the acorn becoming the oak, technical lineages are, instead, tracings or recordings of the becomings of the technical object. Technical lineages, therefore, represent a historiography of the technical object in which it is eventalized.

Technological determinism: This is the fallacious belief that technology fundamentally affects, to the exclusion of other factors, all patterns of social existence and that technological change is the principal source of change in a society. Technological determinism also tends to be reductive in its understanding of technology and typically isolates one technology as being at the root of all social change. Overestimations of the role of the printing press, for example, in the changes that European societies underwent in the Early Modern period depend upon such a misconception. At the same time, it is equally erroneous to adopt cultural determinism as a remedy to technological determinism since it simply applies the same model of causality but in reverse so that technology is simply reduced to an expression of the society. Again, this entails a reductive understanding of technology since it presumes that technologies are themselves neutral and it is how society elects to use them that matters. Ultimately, both approaches can only conceive of the relationship between technology and culture in terms of linear causality.

Transduction: This is the term Simondon uses to refer to the process of individuation. Transduction leads to the becoming of an individuated being in relation to its environment or milieu leading to an individuated being becoming as a metastable state. While individual entities are typically understood as closed and complete, transduction highlights the openness and impermanence of individuated beings.

Transitivity: This refers to the way in which texts depict social action. Transitivity structures realize the representational function of communication by semiotizing represented actions or activities as types of processes (material, behavioral, mental, verbal, relational, and existential), identifying (or obscuring) participants, and elaborating (or not) upon circumstances associated with the action. Detailing the realization of transitivity structures highlights not only the way the actions of actors come to be represented but also the distributions of agency among participants.

Bibliography

Abousnnouga, G. and Machin, D., 2010, Analysing the Language of War Monuments, *Visual Communication*, 9(2), pp. 131–49.
Agamben, G., 2009, *What is an Apparatus?: and Other Essays,* Stanford, CA: Stanford University Press.
Akrich, M., 1992, The De-Scription of Technical Objects, in M. Akrich, W. E. Bijker, and J. Law (eds.), *Shaping Technology/Building Society: Studies in Sociotechnical Change,* Cambridge, MA: MIT Press, pp. 205–24.
Akrich, M. and Latour, B., 1992, Summary of a Convenient Vocabulary for the Semiotics of Human and Nonhuman Assemblies, in W. E. Bijker and J. Law (eds.), *Shaping Technology/Building Society: Studies in Sociotechnical Change,* Cambridge, MA: MIT Press, pp. 259–64.
Anaïs, S., 2013, Genealogy and Critical Discourse Analysis in Conversation: Texts, Discourse, Critique, *Critical Discourse Studies*, 10(2), pp. 123–35.
Armitage, J., 1999, Resisting the Neoliberal Discourse of Technology: The Politics of Cyberculture in the Age of the Virtual Class, *C-Theory*. a068-3/1/1999. Retrieved October 24, 2014, from http://www.ctheory.net/articles.aspx?id=111
Aytes, A., 2013, Return of the Crowds: Mechanical Turk and Neoliberal States of Exception, in T. Scholz (ed.), *Digital Labor: The Internet as Playground and Factory,* New York: Routledge, pp. 79–97.
Baker, III, F. W., 2007, Soldiers Like FCS Test Systems So Much, They Don't Want to Return Them, *U.S. Department of Defense American Forces Information Service,* February 13, 2007.
Baker, P., 2012, Acceptable bias? Using Corpus Linguistics Methods with Critical Discourse Analysis, *Critical Discourse Studies*, 9(3), pp. 247–56.
Bardin, A. and Menegalle, G., 2015, Introduction to Simondon, *Radical Philosophy: Philosophical Journal of the Independent Left*, 189, pp. 15–16.
Barthes, R., 1973, *Mythologies,* translated by A. Lavers, London: Paladin Books.
Barthes, R., 1977, *Image, Music, Text,* translated by S. Heath, New York: Hill and Wang.
Barylick, C., 2006, iRobot's PackBot on the Front Lines, *UPI,* February 23, 2006.
Baudrillard, J., 1981, *For a Critique of the Political Economy of the Sign,* St. Louis, MO: Telos Press.
Baudrillard, J., 1994, *Simulacra and Simulation,* Ann Arbor: University of Michigan Press.
Baudrillard, J., 1996, *The System of Objects,* London; New York: Verso.
Beaune, J., 1989, The Classical Age of Automata: An Impressionistic Survey from the Sixteenth to Nineteenth Century, in M. Feher, R. Naddaff, and N. Tazi (eds.), *Fragments for a History of the Human Body Part One. Part One,* New York: Urzone, pp. 431–80.

BIBLIOGRAPHY

Bender, B., 2006, Panel on Iraq Bombings Grows to $3b Effort, *The Boston Globe,* June 25, 2006.

Benjamin, W., 1968, *Illuminations,* edited by H. Arendt, translated by H. Zohn, New York: Harcourt, Brace & World.

Berger, J., 1972, *Ways of Seeing,* London: British Broadcasting Corp.: Penguin Books.

Bezemer, J. and Jewitt, C., 2009, Social Semiotics, in J.-O. Östman, J. Verschueren, and E. Versluys (eds.), *Handbook of Pragmatics: [2009 Installment],* Amsterdam; Philadelphia: John Benjamins Publishing Company, pp. 1–13.

Bigelow, B., 2006, Cyber-Soldiers May Save Lives: San Diego Lab Developing Robotsfor Battlefield Use, April 19, 2006, p. 1.

Bilton, N., 2014, Can Gadget-free Bedrooms Create the Mood for Happy Relationships? Disruptions, *International New York Times,* December 5, 2014, p. 19.

Bora, B., 2008, Jabil Digs in on Defense, *St. Petersburg Times,* March 10, 2008, p. 1D.

Bower, J. L. and Christensen, C. M., 1995, Disruptive Technologies: Catching the Wave, *Harvard Business Review,* 73(1), pp. 43–53.

Boyd, S., 2006, Robots Saving U.S. Lives in Iraq and Afghanistan, *McClatchy Washington Bureau,* February 23, 2006.

Bradshaw, J. M., 1997, Software Agents, in J. M. Bradshaw (ed.), *An Introduction to Software Agents,* Cambridge, MA: MIT Press, pp. 3–46.

Bray, H., 2007, iRobot Wrests Army Contract from Rival Robotic FX, *The Boston Globe,* December 19, 2007, p. 3.

Burnam-Fink, M., 2012, Drone Wars: Winning the Fight against Terrorists, or Prolonging it? *The Cairo Review of Global Affairs,* 5, pp. 83–93.

Caffentzis, C. G. 1997, Why Machines Cannot Create Value; or, Marx's Theory of Machines, in M. Stack, T. A. Hirschl, and J. Davis (eds.), *Cutting Edge: Technology, Information Capitalism and Social Revolution,* London; New York: Verso.

Campbell, N., 2010, Future Sex: Cyborg Bodies and the Politics of Meaning, *Advertising & Society Review,* 11(1). Retreived April 19, 2015, from DOI: 10.1353/asr.0.0045.

Capek, K. and Selver, P., 1923, *R.U.R. (Rossum's Universal Robots: A Fantastic Melodrama,* Garden City, NY: Doubleday, Page.

Carbone, S., 2014, Tech Addicts Urged to Vote Aye and Look up from iDevices, *The Age,* May 10, 2014, p. 16.

Carey, J., 1997, Afterword: The Culture in Question, in E. S. Munson and C. A. Warren (eds.), *James Carey: A Critical Reader,* Minneapolis: University of Minnesota Press, pp. 308–40.

Carroll, E. A., Czerwinski, M., Roseway, A., Kapoor, A., Johns, P., Rowan, K., and Schraefel, M. C., 2013, Affective Computing and Intelligent Interaction (ACII), 2013 Humaine Association Conference on, *Food and Mood: Just-in-time Support for Emotional Eating,* pp. 252–7.

Christensen, C. M., 1997, *The Innovator's Dilemma: When New Technologies Cause Great Firms to Fail,* Boston, MA: Harvard Business School Press.

Chun, L., 2006, Robotic Technology Lowers Military Risk, *UPI,* June 6, 2006.

Cowan, R. S., 1983, *More Work for Mother: the Ironies of Household Technology from the Open Hearth to the Microwave,* New York: Basic Books.

Craig Kirchoff, A., 2012, Editorial: The Changing Nature of Work, *The Washington Post,* February 6, 2012, p. A16.

Crawford, K., Lingel, J., and Karppi, T., 2015, Our Metrics, Ourselves: A Hundred Years of Self-tracking from the Weight Scale to the Wrist Wearable Device, *European Journal of Cultural Studies*, 18(4–5), pp. 479–96.
Dant, T., 1999, *Material Culture in the Social World: Values, Activities, Lifestyles*, Buckingham; Philadelphia: Open University Press.
Darack, E., 2011, A Brief History of Unmanned Aircraft: From Bomb-bearing Balloons to the Global Hawk, *Air & Space Magazine,* May 17, 2011.
Deleuze, G., 1988, *Foucault,* translated by P. Bové and S. Hand, Minneapolis: University of Minnesota Press.
Deleuze, G., 1992, Postscript on the Societies of Control, *October*, 59, pp. 3–7.
Deleuze, G., 2006, What is a *Dispositif?* in *Two Regimes of Madness: Texts and Interviews 1975-1995,* Semiotext(e); Distributed by MIT Press, Los Angeles, CA; Cambridge, MA, pp. 338–48.
Deleuze, G., Guattari, F., Massumi, B., and University, O. O. M., 2014, *A Thousand Plateaus: Capitalism and Schizophrenia*, Minneapolis; London: University of Minnesota Press.
Djik, T. A., 2009, Critical Discourse Studies: A Sociocognitive Approach, in R. Wodak and M. Meyer (eds.), *Methods of Critical Discourse Analysis*, London: SAGE, pp. 62–86.
Djonov, E. and Zhao, S., 2014, From Multimodal to Critical Multimodal Studies through Popular Culture, in E. Djonov and S. Zhao (eds.), *Critical Multimodal Studies of Popular Discourse*, New York: Routledge, pp. 1–15
Dumouchel, P., 1992, Gilbert Simondon's plea for a Philosophy of Technology, *Inquiry*, 35(3–4), pp. 407–21.
Ellul, J., 1964, *The Technological Society,* New York: Knopf.
Fairclough, N., 1995, *Critical Discourse Analysis: The Critical Study of Language,* New York: Longman.
Fairclough, N., 2013, *Critical Discourse Analysis: The Critical Study of Language.* Longman applied linguistics, 2, revised ed., New York: Routledge.
Fairclough, N. and Wodak, R., 1997, Critical Discourse Analysis, in Teun A. van Dijk (ed.), *Discourse as Social Interaction,* London: SAGE, pp. 258–85.
Fairclough, N., Muldering, J., and Wodak, R., 2011, Critical Discourse Analysis, in T. A. V. Dijk (ed.), *Discourse Studies: A Multidisciplinary Introduction,* London: SAGE, pp. 357–78.
Fleischauer, F., 2006, Unleashing the Robots of War: Redstone's Robotics Chief Strives To Put Technology, Not Soldiers, in Harm's Way, *The Decatur Daily,* June 19, 2006.
Forceville, C., 2013, Relevance Theory as Model for Analysing Visual and Multimodal Communication, in D. Machin (ed.), *Visual Communication,* Berlin: Walter de Gruyter GmbH & Co. KG, pp. 51–70.
Forgione, N., 1999, "The Shadow Only": Shadow and Silhouette in Late Nineteenth-Century Paris, *The Art Bulletin*, 81(3), pp. 490–512.
Foucault, M., 1976, *The Archaeology of Knowledge,* translated by A. Sheridan, New York: Harper & Row.
Foucault, M., 1978, *The History of Sexuality*, New York: Pantheon Books.
Foucault, M., 1990, The Order of Discourse, in R. Young (ed.), *Untying the Text: A Post-structuralist Reader*, London: Routledge, pp. 48–78.
Fox, G., Afghanistan Shura, *Afghanistan Shura | Flickr—Photo Sharing!*. Retrieved April 30, 2015, from https://www.flickr.com/photos/defenceimages/10997023064/in/photolist-hKLBPW-dHcto7-dEFVce-eEHoc9-

dPxXN9-e1jJWx-eEHo9q-e2Vgua-9c3JXu-9jYvCS-aBJtNP-ckz7dm-aBJtQZ-pJc2bu-fj7849-9jYvGS-8sPYNY

Freud, S., 1975, Fetishism, in A. Tyson and A. Freud (eds.), *The Standard Edition of the Complete Psychological Works of Sigmund Freud: Early Psycho-analytic Publications*, London: Hogarth Press, pp. 147–57.

Garamone, J., 2002, From US Civil War to Afghanistan: A Short History of UAVs, *Defense.gov News Article: From U.S. Civil War to Afghanistan: A Short History of UAVs*. Retrieved March 24, 2015, from http://www.defense.gov/news/newsarticle.aspx?id=44164

Garamone, J., 2007, Proposed Cuts Endanger Army's Future Combat System, *American Forces Press Service,* May 15, 2007.

Gariepy, R., 2014, Clearpath Robotics Open Letter against Killer Robots, *Kitchener Post,* August 14, 2014.

Garreau, J., 2007, Bots on the Ground; In the Field of Battle (or even above it), Robots Are a Soldier's Best Friend, *The Washington Post,* May 6, 2007, p. 1.

Gibson, J. J., 1979, *The Ecological Approach to Visual Perception,* Boston: Houghton Mifflin.

Gold, S., 2013, Coming Soon: Personalized Factory Workstations, *Pictures of the Future,* Spring, p. 24.

Gould, S. J., 1989, *Wonderful Life: The Burgess Shale and the Nature of History,* New York: W.W. Norton.

Graham, P., 2001, Space: Irrealis Objects in Technology Policy and their Role in a New Political Economy, *Discourse & Society,* 12(6), pp. 761–88.

Greenberg, A. S., 2005, *Manifest Manhood and the Antebellum American Empire,* Cambridge, UK; New York: Cambridge University Press.

Grove, J., An Insurgency of Things: A Foray into the World of Improvised Explosive Devices, Unpublished Manuscript, https://www.academia.edu/8612626/An_Insurgency_of_Things_A_Foray_into_the_World_of_Improvised_Explosive_Devices. Department of Political Science, University of Hawai'i at Manoa.

Hall, E. T., 1969, *The Hidden Dimension: Man's Use of Space in Public and Private,* London: Bodley Head.

Hall, S., 2001, Foucault: Power, Knowledge and Discourse, in M. Wetherell, S. Yates, S. Taylor, and O. University (eds.), *Discourse Theory and Practice: A Reader,* London; Thousand Oaks, CA: SAGE, pp. 72–81.

Halliday, M. A. K., 1985, *An Introduction to Functional Grammar,* London; Baltimore, MD, USA: E. Arnold.

Hart, C., 2011, Introduction, in C. Hart (ed.), *Critical Discourse Studies in Context and Cognition,* Amsterdam; Philadelphia: John Benjamins Pub. Co., pp. 1–6.

Hart, C., 2014, *Discourse, Grammar and Ideology: Functional and Cognitive Perspectives,* London: Bloomsbury.

Hayles, K., 2002, *Writing Machines,* Cambridge, MA: MIT Press.

Hayward, M. and Thibault, G., 2013, Machinic Milieus: Simondon, John Hart and Mechanology, *Disability History Newsletter,* 9(2), pp. 28–33.

Higbie, T., 2013, Why Do Robots Rebel? The Labor History of a Cultural Icon, *Labor,* 10(1), pp. 99–121.

Hobsbawm, E. J. and Ranger, T. O., 1983, *The Invention of Tradition,* Cambridge [Cambridgeshire]; New York: Cambridge University Press.

Hodgins, J. G., 1858, *The Geography and History of British America, and of the Other Colonies of the Empire; To Which are Added a Sketch of the Various*

Indian Tribes of Canada, and Brief Biographical Notices of Eminent Persons Connected with the History of Canada, 2nd ed., Toronto: MacLear and Co.; James Campbell; and W.F.C. Caverhill.

Hornborg, A., 2011, Animism, Fetishism, and Objectivism as Strategies for Knowing (or not Knowing) the World, *Ethnos,* 71(1), pp. 21–32.

Hume, S., 2014, When Artificial Intelligence becomes Perfectly Natural; What's Next?: Leaps in Knowledge and Tech Innovation are a Double-edged Sword, *Vancouver Sun,* October, 18, p. F5.

Iedema, R., 2001, Resemiotization, *Semiotica,* 137(1/4), pp. 23–39.

Iedema, R., 2003, Multimodality, Resemiotization: Extending the Analysis of Discourse as Multi-semiotic Practice, *Visual Communication,* 2(1), pp. 29–57.

Iedema, R., 2007, On the Multi-modality, Materially and Contingency of Organization Discourse, *Organization Studies,* 28(6), pp. 931–46.

Iliadis, A., 2013, Informational Ontology: The Meaning of Gilbert Simondon's Concept of Individuation, *Communication +1,* 2(5). Retrieved July 8, 2014, from http://scholarworks.umass.edu/cpo/vol2/iss1/5, DOI: 10.7275/R59884XW

Ingold, T., 2012, Toward an Ecology of Materials*, *Annual Review of Anthropology,* 41(1), pp. 427–42.

iRobot, 2007, iRobot Files Lawsuit to Stop Infringement of Patents on Combat-Proven PackBot Robot, *iRobot News Release,* August 20, 2007.

Jewitt, C., 2009a, An Introduction to Multimodality, in C. Jewitt (ed.), *The Routledge Handbook of Multimodal Analysis,* London; New York: Routledge, pp. 14–27.

Jewitt, C., 2009b, Introduction, in C. Jewitt (ed.), *The Routledge Handbook of Multimodal Analysis,* London; New York: Routledge, pp. 1–7.

Jones, J., 2006, Making Robots for the Home or a Battlefield, *The New York Times,* August 12, 2006, p. 3.

Keen, J., 1991, Troops Regret Deaths, Defend Raid; They Place Blame on Saddam, *U.S.A. Today,* February 15, 1991, p. 3.

Klein, A., 2007, The Army's $200 Billion Makeover; March to Modernize Proves Ambitious and Controversial, *The Washington Post,* December 7, 2007, p. 1.

Komarow, S., 2005, Robots Nail Down the Nuts and Bolts of Bomb Disposal, *USA Today,* October 25, 2005, p. 10.

Krasner, J., 2007, Robots Going in Harm's Way: Agile New Devices Save Lives by Disarming or Detonating Roadside Bombs, *The Boston Globe,* March 12, 2007, p. 1.

Kress, G. R., 2009, What is a Mode? in C. Jewitt (ed.), *The Routledge Handbook of Multimodal Analysis,* London; New York: Routledge, pp. 54–67.

Kress, G. R., 2010, *Multimodality: A Social Semiotic Approach to Contemporary Communication,* London; New York: Routledge.

Kress, G. R. and Adami, E., 2010, The Social Semiotics of Convergent Mobile Devices: New Forms of Composition and the Transformation of *habitus,* in *Multimodality: A Social Semiotic Approach to Contemporary Communication,* London; New York: Routledge, pp. 184–96.

Kress, G. R. and van Leeuwen, T., 2001, *Multimodal Discourse: The Modes and Media of Contemporary Communication,* New York: Arnold; London: Oxford University Press.

Kress, G. R. and van Leeuwen, T., 2006, *Reading Images: The Grammar of Visual Design,* 2nd ed., London; New York: Routledge.

Lakoff, G. and Johnson, M., 1980, *Metaphors We Live by,* Chicago: University of Chicago Press.

LaSalle, L., 2012, "Internet of Things" Shares Data without Your Help: Potential for Smart Devices is "boundless," says Waterloo Chip Developer, *The Record,* April 28, 2012, p. C1.

Lassen, I., 2006, Is the Press Release a Genre? A study of form and content, *Discourse Studies,* 8(4), pp. 503–30.

Latour, B., 1986, Visualization and Cognition: Drawing Things Together, in E. Long, L. Hargens, R. A. Jones, A. Pickering, and H. Kuklick (eds.), *Knowledge and Society: Studies in the Sociology of Culture Past and Present: A Research Annual,* Greenwich, CT; London: JAI Press Distributed by JAICON Press, pp. 1–40.

Latour, B., 1987, *Science in Action: How to Follow Scientists and Engineers through Society,* Cambridge, MA: Harvard University Press.

Latour, B., 1990, Technology is Society made Durable, *The Sociological Review,* 38(S1), pp. 103–31.

Latour, B., 1992, Where Are the Missing Masses? The Sociology of a Few Mundane Artifacts, in W. Bijker and J. Law (eds.), *Shaping Technology/Building Society: Studies in Sociotechnical Change,* Cambridge, MA: MIT Press, pp. 225–58.

Latour, B., 2004, Why Has Critique Run out of Steam? From Matters of Fact to Matters of Concern, *Critical Inquiry,* 30(2), pp. 225–48.

Law, J., 1991, *A Sociology of Monsters: Essays on Power, Technology, and Domination,* London; New York: Routledge.

Lazzarato, M., 1996, Immaterial Labour, in M. Hardt and P. Virno (eds.), *Radical Thought in Italy: A Potential Politics,* Minneapolis, MN: University of Minnesota Press, pp. 132–46.

Lefebvre, A., 2011, The Individuation of Nature in Gilbert Simondon's Philosophy and the Problematic Nature of the Technological Object, *Techné: Research in Philosophy and Technology,* 15(1), pp. 1–15.

Lemke, J. L., 1993, Discourse, Dynamics, and Social Change, *Cultural Dynamics,* 6(1), pp. 243–75.

LeVine, P. and Scollon, R., 2004, *Discourse and Technology: Multimodal Discourse Analysis,* Washington, DC: Georgetown University Press.

Lister, M., Dovey, J., Giddings, S., Grant, I., and Kelly, K., 2008, *New Media: A Critical Introduction,* revised ed., London: Routledge.

Lucchese, F. D., 2009, Monstrous Individuations: Deleuze, Simondon, and Relational Ontology, *Differences,* 20(2–3), pp. 179–93.

Lyotard, J.-F., Bennington, G., and Massumi, B., 1984, *The Postmodern Condition: A Report on Knowledge,* Minneapolis, MN: University of Minnesota Press.

Machin, D., 2004, Building the World's Visual Language: The Increasing Global Importance of Image Banks in Corporate Media, *Visual Communication,* 3(3), pp. 316–36.

Machin, D., 2010, *Analysing Popular Music: Image, Sound, Text,* Los Angeles: SAGE.

Machin, D., 2011, *Introduction to Multimodal Analysis,* New York: Bloomsbury Academic.

Machin, D., 2013, What is Multimodal Critical Discourse Studies? *Critical Discourse Studies,* 10(4), pp. 347–55.

Machin, D. and Mayr, A., 2012, *How to do Critical Discourse Analysis: A Multimodal Introduction,* Los Angeles: SAGE.

Machin, D. and van Leeuwen, T., 2005, Computer Games as Political Discourse: The Case of, *Journal of Language and Politics,* 4(1), pp. 119–41.

Mackenzie, A., 2002, *Transductions: Bodies and Machines at Speed,* London; New York: Continuum.

Mackenzie, A., 2005, Problematising the Technological: The Object as Event? *Social Epistemology,* 19(4), pp. 381–99.

Martin, J. R. and Veel, R., 1998, *Reading Science: Critical and Functional Perspectives on Discourses of Science,* London; New York: Routledge.

Marx, K., 2005a, The Labour-Process and the Process of Producing Surplus-Value, in *Capital: A Critique of Political Economy,* www.marxists.org.

Marx, K., 2005b, Section 4—*The Fetishism of Commodities and the Secret thereof,* in *Capital: A Critique of Political Economy,* www.marxists.org.

Masuzawa, T., 2000, Troubles with Materiality: The Ghost of Fetishism in the Nineteenth Century, *Comparative Studies in Society and History,* 42(02), pp. 242–67.

Maunders, S., 1859, A Compendius Universal Gazetteer; Derived from the Latest and Best Authorities, and Condensed into the Smallest Space Possible, Compatible with Real utility; Garnished with Proverbs of all Nations, in B. B. Woodward (ed.), *The Treasury of Knowledge and Library Reference The Treasury of Knowledge and Library Reference,* London: Longman, Brown, Green, Longmans, and Roberts.

McDonald, S. N., 2014, Missouri Highway Patrol Capt. Ron Johnson is not a Gang Member. He's just a Kappa., *Missouri Highway Patrol Capt. Ron Johnson is not a Gang Member. He's just a Kappa.—The Washington Post,* August 20, 2014.

McHoul, A. W. and Grace, W., 2002, *A Foucault Primer Discourse, Power, and the Subject,* London; New York: Routledge.

McKenna, B. J. and Graham, P., 2000, Technocratic Discourse: A Primer, *Journal of Technical Writing and Communication,* 30(3), pp. 223–52.

Mills, S., 2011, FCJ-127 Concrete Software: Simondon's Mechanology and the Techno-social, *The Fibreculture Journal* (issue 18 2011: Trans), Retrieved July 8, 2014, from http://eighteen.fibreculturejournal.org/2011/10/09/fcj-127-concrete-software-simondon%e2%80%99s-mechanology-and-the-techno-social/

Morozov, E., 2014, *To Save Everything, Click here: the Folly of Technological Solutionism,* New York: PublicAffairs.

Morrison, B. and Eisler, P., 2007, Destroy or Investigate? A Commander's Choice; New "Blow and Go" Policy Complicates Push to Track Bombmakers, *USA Today,* November 7, 2007, p. 1.

mullen.com, 2011, *iRobot. Do you?.* Retrieved February 14, 2015, from http://www.mullen.com/work/irobot-do-you/?t=0

Myers, G., 1996, Out of the Laboratory and Down to the Bay Writing in Science and Technology Studies, *Written Communication,* 13(1), pp. 5–43.

Myers, G., 2003, Discourse Studies of Scientific Popularization: Questioning the Boundaries, *Discourse Studies,* 5(2), pp. 265–79.

Nisbet, R. A., 1980, *History of the Idea of Progress,* New York: Basic Books.

Noble, D. F., 1983, Present Tense Technology, *Democracy,* 3(2), pp. 8–24.

Noble, D. F., 1984, *Forces of Production: A Social History of Industrial Automation,* New York: Knopf.

Nye, D. E., 1994, *American Technological Sublime,* Cambridge, MA: MIT Press.
Penley, C. and Ross, A., 1991, *Technoculture,* Minneapolis: University of Minnesota Press.
Person of the Year | TIME, *Time Magazine.* Retrieved December, 2013, from http://poy.time.com/
Piazza, J., 2006, EOD Team Calm through IED Storm, *News Blaze,* August 5, 2006.
Pontin, B. J., 2007, Artificial Intelligence, With Help From the Humans, *The New York Times,* March 25, 2007.
Rajchman, J., 1988, Foucault's Art of Seeing, *October,* 44, pp. 88–117.
Reisigi, M. and Wodak, R., 2009, The Discourse-Historical Approach, in R. Wodak and M. Meyer (eds.), *Methods of Critical Discourse Analysis,* 2nd ed., Thousand Oaks, CA: Sage, pp. 87–121.
Robinson, P., 2004, Researching US Media-State Relations and Twenty-First Century Wars, in S. Allan and B. Zelizer (eds.), *Reporting War: Journalism in Wartime,* London; New York: Routledge, pp. 96–112.
Roderick, I., 2010, Considering the Fetish Value of EOD Robots: How Robots Save Lives and Sell War, *International Journal of Cultural Studies,* 13(3), pp. 235–53.
Roderick, I., 2013, Representing Robots as Living Labour in Advertisements: The New Discourse of Worker–Employer Power Relations, *Critical Discourse Studies,* 10(4), pp. 392–405.
Rosenberger, R., 1997, When and how did the Metaphor of the Computer "virus" Arise? *ScientificAmerican.com,* September 2, 1997.
SAIC Produces Future Force Company Commander (F2C2), 2006.
Sanchez, S., 2005, *Operational Test Command Public Affairs Office,* January 28, 2005.
Saussure, Ferdinand de, 1959, *Course in General Linguistics,* New York: Philosophical Library.
Schafer, R. J., 2003, Robotics to Play Major Role in Future Warfighting, *United States Joint Forces Command,* July 29, 2003.
Schumpeter, J. A., 1950, *Capitalism, Socialism, and Democracy,* New York: Harper.
Schutz, A., 1967, *The Phenomenology of the Social World,* Evanston, IL: Northwestern University Press.
Segall, K., 2012, *Insanely Simple: The Obsession that Drives Apple's Success,* New York, NY: Portfolio.
Sestak, Rep. J D-Pa. (7th CD) 2008, 'Rep. Sestak Tours Chatten Associates', West Conshohocken, PA: US Federal News, January 30.
Shachtman, N., 2005, The Baghdad Bomb Squad, *WIRED,* 2005.
Sharma, P., Cheikh, F. A., and Hardeberg, J. Y., 2009, Face Saliency in various Human Visual Saliency Models, *IEEE Xplore,* pp. 327–32. Retreived November 3, 2014, from 10.1109/ISPA.2009.5297732
Shaw, D. B., 2008, *Technoculture: The Key Concepts,* Oxford: Berg.
Shaw, G., 2013, Is there a Line between Company Time and Personal Time?; Technology makes workers available 24/7, but that doesn't make it right, *The Vancouver Sun (British Columbia),* April 13, 2013, p. C1.
Shaw, M., 2002, Risk-transfer Militarism, Small Massacres and the Historic Legitimacy of War, *International Relations,* 16(3), pp. 343–59.

Shaw, M., 2005, *The New Western Way of War: Risk-transfer War and its Crisis in Iraq,* Cambridge: Polity.
Silverberg, D., The Remote Controlled Military and the Future of Warfare, *Digital Journal.* Retrieved February 2, 2006.
Simondon, G., 1980, *On the Mode of Existence of Technical Objects,* translated by N. Mellamphy, London, ON: University of Western Ontario.
Simondon, G., 1992, The Genesis of the Individual, in J. Crary and S. Kwinter (eds.), *Incorporations,* Cambridge, MA: Zone Books, pp. 296–319.
Skinner, N., 2013, Work is where the Home is; Home-Life Balance, *The Age (Melbourne, Australia),* February 23, 2013, p. 3.
Slack, J. D. and Wise, J. M., 2005, *Culture + Technology: A Primer,* New York: Peter Lang.
Speckman, S., 2006, Better Technology Helps Troops Disarm IEDs, *Deseret Morning News,* December 9, 2006.
Spires, S., 2006, UAVs Take Off, *Huntsville Times,* December 10, 2006, p. 1.
Stenglin, M., 2009, Space and Communication in Exhibitions: Unravelling the Nexus, in C. Jewitt (ed.), *The Routledge Handbook of Multimodal Analysis,* London; New York: Routledge, pp. 272–83.
Thacker, E., 2004, *Biomedia,* Minneapolis, MN: University of Minnesota Press.
Thompson, J. B., 1984, *Studies in the Theory of Ideology,* Berkeley: University of California Press.
Trevino, Sgt A., 2007, Saving Robots to Save Battlefield Lives, *Department of Defense US Army Releases,* June 5, 2007.
Tronti, M., 1966, *Operai e Captiale (Workers and Captial),* Turin: Einaudi.
Turk, G., 2014, Look Up, video, viewed April 20, 2015, https://www.youtube.com/watch?v=Z7dLU6fk9QY
US Army, 2007a, Army shows Congress FCS "spin-out" technologies, *Department of Defense US Army Releases,* September 28, 2007.
US Army, 2007b, Robotic Contract Awarded to Save Lives, Limbs in Theater Despite Set-Back, *Program Executive Office for Simulation, Training & Instrumentation,* December 20, 2007.
US Army, 2008, Sen. Carl Levin and Tardec Director Encourage University Students to Pursue Robotics Education at 16th Annual IGVC, *Department of Defense US Army Releases,* June 16, 2008.
van Dijk, T. A., 2001, Multidisciplinary CDA: A Plea for Diversity, in R. Wodak and M. Meyer (eds.), *Methods of Critical Discourse Analysis,* London; Thousand Oaks, CA: SAGE, pp. 95–118.
van Dijk, T. A., 2008, *Discourse and Context: A Socio-cognitive Approach,* Cambridge; New York: Cambridge University Press.
van Leeuwen, T., 1995, Representing Social Action, *Discourse & Society,* 6(1), pp. 81–106.
van Leeuwen, T., 1996, The Representation of Social Actors, in C. R. Caldas-Coulthard and M. Coulthard (eds.), *Texts and Practices: Readings in Critical Discourse Analysis,* London; New York: Routledge, pp. 32–70.
van Leeuwen, T., 2004, Ten Reasons Why Linguists Should Pay attention to Visual Communication, in P. LeVine and R. Scollon (eds.), *Discourse and Technology: Multimodal Discourse Analysis,* Washington, DC: Georgetown University Press, pp. 7–19.
van Leeuwen, T., 2005, *Introducing Social Semiotics,* London; New York: Routledge.

van Leeuwen, T., 2006, Towards a Semiotics of Typography, *Information Design Journal*, 14(2), pp. 139–55.
van Leeuwen, T., 2008, *Discourse and Practice: New Tools for Critical Discourse Analysis,* Oxford; New York: Oxford University Press.
van Leeuwen, T., 2012, Critical Analysis of Multimodal Discourse, in C. Chapelle (ed.), *The Encyclopedia of Applied Linguistics,* New York: John Wiley and Sons, pp. 1–6.
Vološinov, V. N., 1973, *Marxism and the Philosophy of Language: Translated by Ladislav Matejka and I. R. Titunik,* New York: Seminar Press.
de Vries, M. J., 2008, Gilbert Simondon and the Dual Nature of Technical Artifacts, *Techné: Research in Philosophy and Technology*, 12(1), pp. 23–35.
Walt Disney's Carousel of Progress | Magic Kingdom Attractions | Walt Disney World Resort. Retrieved April 29, 2015, from https://disneyworld.disney.go.com/attractions/magic-kingdom/walt-disney-carousel-of-progress/
Weaver, R. M., 1953, *The Ethics of Rhetoric,* Chicago: H. Regnery Co.
Weber, S., 2012, Keep it Simple!: Mastering Complexity | User-Friendlieness, *Pictures of the Future,* 2012, pp. 58–59.
White, P., 2000, Death, Disruption and the Moral Order: The Narrative Impulse in Mass-media "Hard News" Reporting, in J. R. Martin and F. Christie (eds.), *Genre and Institutions Social Processes in the Workplace and School,* London; New York: Continuum, pp. 101–33.
Williams, R., 2003, *Television: Technology and Cultural Form,* Routledge; New York: Routledge Classics Edition.
Williams, R., 1985, *Keywords a Vocabulary of Culture and Society,* Oxford: Oxford University Press.
Winner, L., 1977, *Autonomous Technology: Technics-out-of-control as a Theme in Political Thought,* Cambridge, MA: MIT Press.
Winner, L., 1980, Do Artifacts Have Politics? *Daedalus,* 109(1), pp. 121–36.
Wodak, R., 2011, Critical Discourse Analysis, in K. Hyland and B. Paltridge (eds.), *Continuum Companion to Discourse Analysis,* London; New York: Continuum, pp. 38–53.
Wodak, R. and Meyer, M., 2009, Critical Discourse Analysis: History, Agenda, Theory, and Methodology, in R. Wodak and M. Meyer (eds.), *Methods of Critical Discourse Analysis,* 2nd ed., Thousand Oaks, CA: Sage, pp. 1–33.
Zhao, S. and van Leeuwen, T., 2014, Understanding Semiotic Technology in University Classrooms: A social Semiotic Approach to PowerPoint-Assisted Cultural Studies Lectures, *Classroom Discourse,* 5(1), pp. 71–90.
Zhao, S., Djonov, E., and van Leeuwen, T., 2014, Semiotic Technology and Practice: A Multimodal Social Semiotic Approach to PowerPoint, *Text & Talk,* 34(3), pp. 349–75.
Zombie Studios Inc., 2002, *America's Army Operations,* Seattle, WA: Computer Game.
Zombie Studios Inc., 2005, *Future Force Company Commander,* Seattle, WA: Computer Game.

Index

Actor-Network Theory 5, 50, 151, *see also* resemiotization
Adobe, *Marketing Cloud* 179–80, 183–6
affordance 5, 62–3; and email text 72; modal 42–5, 52; Paths 105; and racism 37–8; of semiotic resource 37–8; and smart phones 128–9; and technical lineage 205
Agamben, Giorgio 25–6, 193, *see also* technology as apparatus
agency 19, 35, 73, 79, 183, 206; and fetishism 139; and robots 152–61, 182–4, 190; and technological determinism 117, 119, 123, 129–30, 134, 138; and transitivity 43–4, 206
American Progress 97–9, 101–4, 114–15, 190, 193–4
anthropomorphism 66, 68, 71–2
artificial intelligence 66–7, 69, 152, 156, 169
Asimo robot 94, 96–7, 111–14
associated milieu 16–21, 27, 115, 138; and computer mediated technology (CMC) 123, 125; and concretization 201; and drones 21, 24; and EOD fetish 164–7; and iPhone 130; and technical lineage 205; technological determinism 117
automation 169, 179–81, 190; as social labor 180, 183, 189, *see also* Adobe's *Marketing Cloud*; iRobot's *Robot*
automobile 17, 19–20

Barthes, Roland 41, 49, 60, 63–4, *see also* connotation
body, and technology 2–3, 51, 71–2, 174

capitalism 103–4, 188–90, 195
Carousel of Progress 7, 94, 104–11, 115, 194
Coca-Cola (*Social Media Guard*, commercial) 123, 138
color 13, 62, 64, 84, 91, 107; and modality 84; and salience 47–8
concretization 16–19; and Predator drone 23; and techno-fetishism 167
conduit metaphor 42, 87–8
connotation 11–13, 60–4, 72, 99, 158, 163; and denotation 60–1; *vs.* metaphor 203
consumer 34–5, 40, 76–8; as activated or passivated actors 76–8; culture 103–4, 108–11, 170, 174, 179, 191, 194–5
Critical Discourse Analysis (CDA) 3–6, 29–32, 53–9; and metaphor 68; and social action 78, *see also* MCDA
Critical Discourse Studies (CDS) 3–7, 29–31, 51–2, 197; critique as praxis 53–60, 92, *see also* de-naturalizing; MCDA

Deloitte, *Digital Disruption – Short fuse, big bang?* (promotional video) 134–7
de-naturalizing 57–8, 92
discourse: as constitutive of technical object 197; definition 29–35, 50–1; racist 38; resemiotization 50–1; as resource 33, *see also* CDA; CDS; MCDA; technological determinism; technological fetishism; technology as progress

INDEX

drone, as technical object 20–4, 86, 163; Predator drone 20, 22–4; Reaper drone 20, 24
dualism 1, 10–13, 19, 202

Explosive Ordnance Disposal (EOD) 8, 140, 152, 158–64; fetish 164–7

fitness tracker 2–3, 51
Foucault 7, 25, 31–2, 51, 53, 57–8, 92
Future Force Company Commander (F2C2) 8, 140–8

Gender 10–13, 26, 43–8, 97–101, 108–10
GE (Profile ad) 10–13

Hallidayan linguistics 4, 37, 48, *see also* image acts
heterogeneity 24, 30, 50, 197
Home computing 32–5, 126
Honda, *see* Asimo robot
hylomorphism 6, 11, 13, 15, 27, 108, 138

image 5, 11–13, 29, 39–50, 62–4, 80–6, 89–92, 99–100, 204; acts 46–7, 80, 202; interaction analysis 79–83; *vs.* language 44–6, 83
Improvised Explosive Devices (IEDs) 16, 153, 157, 162, 164–6
individuation 10, 14–20, 26–7, 130, 166–7, 196–7, 205; and relationality 204; and transduction 205
InnerVision 173–4
interaction analysis 62, 79–83
Internet of Things 65–7
iPhone 7, 118, 121, 125–30
iPod 39–41, 48
iRobot 159, 174–8; *Robot* (commercial) 179–80, 181–4, 186; *Robot Dance* (commercial) 176–8, *see also* automation; technological consumption; technological (dis)satisfaction

Kress, Gunther: conceptual processes 99; discourse 32–6; exhibitionary spaces 104; hardware and software affordances 128–30; ideology of choice 170–2; images 44–6, 80, 102; layout 89–90; modality 83–5; multimodality 48–9, 61; semiotic mode 41–3, 45, 204; semiotic resources 36–8; salience 47–8; visual representations 10–11

labor 60, 96; crowdsourced 38, 69; domestic 44–5, 108, 110; Human Intelligence Tasks (micro labor) 69; invisible 167; labor savings 8, 170, 179; micro labor 69; precarious labor 179, 180, 188–9; robot as laborer 60, 161, 163–4, 179–84, 187–91; social labor 149–50, 194
language 5, 42–6, 48–52, 64–7; *vs.* image 29, 44–8; and interaction analysis 79–80, 83
Latour, Bruno 1, 5, 7, 14, 15, 19, 50, 53, 55–6, 61, 150, 196; Akrich, Madeline and 151, 196
layout 89–92, 175
logocentrism 29
Look Up (Gary Turk, video) 120–3, 168, 194–5

Machin, David, and CDS 4, 32; de-naturalizing 92; description 59–61, 73; images 11, 44, 46, 49, 64, 86, 88–9; modality 83; music/sound 113, 121, 146–7, 187; representing social actors 65, 73, 76; toolkit approach 7, 54, 57, 61, 92
Marketing Could (commercial) 179–84, *see also* automation, technological (dis)satisfaction
Massumi, Brian 14, 184
materiality 42, 50–2, 151, 196–7
Mechanical Turk, the 69
metafunction 88, 104–5; and mode 41–8
metaphor 62–3, 122, 129, 131; computer security as immune system 67–9; e-waste as bodily waste 71–3; fetish 150–1; and Mechanical Turk 123; process 131, 135–7; robot as dog 159; silhouette as inner self 39–40; technology as

double edged sword 69–71, *see also* conduit metaphor
Miele (vacuum ad) 45–8
modal, affordance 42–5; resource 42
modality 83–7, 121, 146
Multi-Function, Agile Remote-Controlled Robot (MARCBOT) 153–9
Multimodal 6–7, 29–30, 41, 48–50, 59, 196–7; discourse 4–5, 48–52; text 48–50, 52, 60–1, 64, 72, 83, 89–90, 121, 133
Multimodal Critical Discourse Analysis (MCDA) 6, 29, 48–52, 54–62, 92
Multimodal Discourse Analysis (MDA) 6, 59
music 106, 113, 121, 141–8, 177, 180, 186–7; normative claims 5

Nike, *Fuel* 51
Noble, David 93, 158, 161, 169, 194

power 118, 129, 148; and discourse 51; and fetishized warfare 158, 161–3, 165, 194; making relations visible 55, 57–8, 61–2, 73, 92; and progress 93–4, 114, 193–4; and technological determinism 118, 138, *see also* disruptive technologies; imperialism; labor; sublime
privacy, loss of 2, 51, 63, 65, 67, 174
Project Tomorrow 8, 170–3
Prominence 105–8

race, black 5, 37–8; East Asian 11–13; indigenous 94–5, 148; middle eastern 75–7; Native American 97–101; white 11–13, 38, 98–101, 108
relationality 14, 166; within CDS approach 197
representation: and CDS 55; backgrounding 73–5, 83, 154, 156; and modality 83–6; of represented and interactive participants 45–6, 51–2; of social actions 43–4, 78–9; of social actors 43–3, 52–3, 73–8; suppression 73–4, 157
resemiotization 48–51, 196–7

robots 7–8, 60, 96–7, 111–14, 132, 140, 152–67, 174–8, 181, 188–91; as affective bodies 111–13, 184–8; as agents 152–6, 183–4, 190; as double-edged sword 70–2; as fetish 152–4, 158, 160–1, 164–7; humans as 122, 176–7, 181; as objects of affection 159–61; as social actors 183, 187; as social machines 189; as workers 178–83, *see also* Asimo robot; Explosive Ordnance Disposal (EOD); iRobot; Multi-Function Agile Remote-Controlled Robot (MARCBOT)
Roomba 174–8

salience 47–8, 89–90, 105, 125, 146
semiosis 3, 33, 42, 48, 51, 60–3, 73, 196–7
semiotic mode 5–7, 30, 41–4, 48–52, 59–60, 71, 83, 88, 200, 202, 204, *see also* images; materiality; music; speech; typography
semiotic resource 4, 6, 30, 35–45, 50, 52, 55, 61–4, 71, 78–80, 100, 103, 105, 113, 182, 184, 187, 196, 199–201, 203–5, *see also* color; connotation; interaction analysis; layout; metaphor; modality; transitivity analysis; typography
semiotics 3, 6, 29, 36, 204; Akrich and Latour 196; Iedema 196–7; Parisian 48, 200, 204; Saussurean 36; social 6, 29–30, 39, 51, 62–3, 200, 203, *see also* resemiotization
Science Applications International Corporation (SAIC) 91
Siemens, *Coming Soon: Personalized Factory Workstations* 131–5, 173; *Pictures of the Future* 131, 137, 195
Simondon: associated milieu 17–20, 27, 200; concretization 200–1; individuation 14–17, 26–7; relationality 204; relationship between technology and culture 6, 10, 19–20, 27; technical lineage 17, 205; transduction 196, 205

INDEX

Slack, Jennifer Daryl and J. MacGregor Wise   5, 9, 24, 118, 119, 170, 178–9
smart bra   26, 85
Smart devices   65–6, 70, 77, 110, 122, 128–31, 134, 172, see also iPhone; smart bra
SMART Technologies   80–3
Social Construction of Technology (SCOT)   195–6
speech   6, 30, 41, 44, 46, 58, 88–9, 113, 156–8, 174, 183, 186, 202
sublime   8, 139–41, 170; as religious experience 140–8, 178

technical lineage   17–24, 27, 115
technocratic discourse   2, 130–7
technoculture   1–2, 7, 19, 53, 119, 195, see also dualism; hylomorphism
technological determinism   1–2, 4, 7–8, 139–40, 167, 193–7; and agency 129–30; and causality 118–19; and technocratic discourse 130–7; and technological solutionism 125–30; and utopia 131, 133–4, 137, 169, see also Coca-Cola; iPhone; Look Up; Siemen's
technological (dis)satisfaction   4–5, 179, 190–1; as ambivalence 179, 181; as convenience 178, 179, 190–1; Noble on 169; as sublime 178; technological consumption 170, see also Slack and Wise
technological fetishism   4, 7–8, 139–40, 167; as a concept 148–52; EOD 164–7; fetish value of robots 161–3; robot 152–4, 159–61, 164–7; sublime 140–8, 193–6
technological solutionism   59, 125–30
technology: and ambivalence   169–70, 179, 181, 191, 194; as destruction 119–24; and identity 61, 176; impact 1, 5, 18, 119, 134, 174; our relationship to 178, 181, 190–1; and revolution 118–19, 124–5, 181

technology, disruptive   2, 7, 118, 125, 134–7, see also Deloitte
technology as apparatus   9–10, 24–7, 193, 195; and technological determinism 119
technology as progress   4–5, 7–8, 18–19, 67, 103, 111–14, 125, 169, 193–5; as betterment 93–4, 100–1; and democratization of consumption 20, 104, 194; and dissatisfaction 169–70; imperialism 94–5, 97–103; linear 20–4, 95–6, 103, 110–11, 113–14; moral and social superiority 7, 104; as mundane 103–10, 115, 125, 140; and nationalism 100, 103, 114; as natural evolution 94–6; religious connotations 93, 115; shift to personal convenience 115, 131–4, 170, 190–1, 195; and the sublime 140, 178, see also American Progress; Asimo; Carousel of Progress
Time magazine   32–3, 35; Life Books 96
Tomorrowland, see Carousel of Progress
transduction   14–16, 19, 196, 202
transitivity   43, 62, 73, 78, 124, 154
Turk, Gary   120–3, 194–5
typeface   40, 107, 174–6; normative claims 5; as semiotic resource 86–9
typography   62, 72, 86–9; as semiotic mode 88

United States Department of Defense's (DoD) Business Transformation Agency (press release)   152–9
Unmanned Aerial Vehicles (UAV), see drones

van Leeuwen, Theo: agency   73–4, 78, 154, 156–7; connotation 63–4; discourse 32–6; images 44–6, 80, 102; layout 89–90; metaphor 68–9, 71; modality 83–5; multimodality 49, 59, 61; semiotic modes 42–3,

45, 204; semiotic resources 36–8; symbolic processes 99; transitivity 73; typography 86–8; visual representations 10–11

Walt Disney, *see Carousel of Progress; Project Tomorrow*

warfare: Afghanistan 21, 75–8, 152, 161, 163, 165–6; American Civil War 20–2; Cold War 23, 103–4, 194; Desert Storm 22; Enduring Freedom 74; Gulf War 22–3, 74–5; Iraq 22–3, 74–5, 152–3, 157, 159–61, 163, 165–6; Vietnam War 22; World War II 20–3, 194, *see also* drones; Explosive Ordnance Disposal; *Future Force Company Commander*; Improvised Explosive Device

Winner, Langdon 5, 118, 120, 181